P9-CKS-157

More Advance Praise for *Meathooked*

"*Meathooked* is a fascinating, and often surprising, exploration of the human carnivore. At every step of the way, the story of meat eating is more interesting and more complicated than you'd expect. Zaraska provides convincing, and provocative, evidence that we eat meat today for reasons that few people would imagine. It has less to do with nutrition than with culture, marketing, taste and habit. This is a book that every meat eater should read."

—Christopher Leonard, Fellow at the New America Foundation and author of *The Meat Racket*

"We know producing and consuming it is terrible for us, the planet, and billions of farm animals, so what keeps people hooked on meat? Marta Zaraska's fascinating *Meathooked* provides a lively, compelling look at the many reasons humans are addicted to animal protein. Whether you're a vegan, a hardcore meat-lover, or somewhere in between, this book will help you better understand why you and your loved ones eat what you do."

—David Robinson Simon, author of *Meatonomics: How the Rigged Economics of Meat and Dairy Make You Consume Too Much—and How to Eat Better, Live Longer, and Spend Smarter*

"Sometimes the secret is asking the right questions. By examining the positive and negative history of meat rather than vegetarianism, Marta Zaraska leads us to a thoughtful and broad array of issues. *Meathooked* is a book people need to read."

—Mark Kurlansky, bestselling author of *Salt* and *Cod*

MEATHOOKED

THE HISTORY AND SCIENCE OF OUR 2.5-MILLION-YEAR OBSESSION WITH MEAT

MARTA ZARASKA

BASIC BOOKS
A MEMBER OF THE PERSEUS BOOKS GROUP
NEW YORK

Published by Basic Books,
A Member of the Perseus Books Group

Books published by Basic Books are available at special discounts for bulk purchases in the United States by corporations, institutions, and other organizations. For more information, please contact the Special Markets Department at the Perseus Books Group, 2300 Chestnut Street, Suite 200, Philadelphia, PA 19103, or call (800) 810-4145, ext. 5000, or e-mail special.markets@perseusbooks.com.

Designed by Jack Lenzo

Library of Congress Cataloging-in-Publication Data
Zaraska, Marta, author.
Meathooked : the history and science of our 2.5-million-year obsession with meat / Marta Zaraska.
pages cm
Includes bibliographical references and index.
ISBN 978-0-465-03662-2 (hardcover) — ISBN 978-0-465-09872-9 (e-book)
1. Meat—Psychological aspects. 2. Meat industry and trade. 3. Food preferences. 4. Meat animals. 5. Diet. I. Title.
TX371.Z37 2016
641.3'6—dc23

2015027674

10 9 8 7 6 5 4 3 2 1

For Maciek and Ellie

CONTENTS

Introduction

In the summer of 2009, my mother decided to go vegetarian. She had been living among vegetarians for years—both her husband (my stepfather) and his son (my stepbrother) eschew meat. Being a good Polish wife, she would cook them plant-based dinners every night and a separate one, containing meat, for herself. No one pressured her to change her diet, and she didn't seem to mind the additional work. But in 2009 she stumbled upon an article on the health risks of eating meat. The data it quoted, which came from a study of over half a million people, was alarming: high intake of red meat increases a woman's risk of premature death due to heart disease by 50 percent and due to cancer by 20 percent. That, my mother thought, was disturbing. She didn't want to clog her arteries with LDL cholesterol (the bad one) and damage her cells with polycyclic aromatic hydrocarbons (carcinogenic substances that may form during the cooking of meat). She pledged to take better care of herself. She was done with meat, she told us.

My mother's resolve lasted about a fortnight. Then the juicy hams and the creamy pâtés crept back into her fridge. Since that summer she has tried giving up meat several times more, but it has never worked out. Her efforts invariably remind me of my husband's never-ending quest to quit smoking. At some point, when I asked my mother what happened to her vegetarianism, she just shrugged. "I like meat, I eat it, end of story."

But for me, it was only a beginning. Several questions were starting to form in my head, questions about our relationship with meat: What is it about animal protein that makes us crave it? What makes it so hard to give up? And if consuming meat is truly unhealthy for us, why didn't evolution turn us all into vegetarians in the first place?

Two years later in early 2011, I was sitting in Eight Treasures restaurant in Singapore, overlooking the hustle and bustle of Chinatown. Through an open window the smells of incense and frangipani flowers drifted in from a nearby Buddhist temple. The world outside simmered with noise, yet the restaurant was peaceful. By then I'd been living in Singapore for over two months and was slowly getting accustomed to the culture. But in Eight Treasures I was in for another eye-opener. This was supposed to be a vegetarian restaurant, yet the menu featured exclusively meat dishes: mutton curry, suckling pig, Peking duck, even the notoriously environmentally unfriendly shark's fin soup. Confused, I called the waiter. "Do you have any vegetarian dishes at all?" I asked. He looked at me as if I were not exactly sane. "These are *all* vegetarian dishes," he said. Me: "You mean these pork ribs are not made of, well, pork?" The waiter: "Everything is fake meat only."

Ah. Here were the key words: *fake meat*. Soy- or gluten-based mixtures, sometimes flavored with petroleum derivatives. It didn't sound very encouraging, but I took the plunge and ordered "pork" ribs. And they were delicious. They looked like meat; they had the texture of meat; they even tasted like meat. I'm still not 100 percent sure they weren't actually made of meat. Perhaps the cooks at Eight Treasures just serve the vegetarians animal protein and fool them into believing it is soy. But what it really made me wonder was this: Why does such an oddity like fake meat exist at all? We don't concoct fake nuts for those who are allergic, nor are there fake carrots for the strict Jains, who avoid root vegetables (they believe pulling them out of the ground is gross violence). So why bother with fake meat? Are we so addicted to animal protein that we'd rather eat a meat-substitute curry loaded with chemicals than just enjoy a simple dish of curried vegetables? What's in the taste of meat or in its social and cultural appeal so that even lifelong vegetarians can't completely give it up?

Today, my mother still eats meat. She is even known to enjoy such Polish delicacies as *kaszanka* (a sausage made of pig's blood and lungs) and *wątróbka* (seared cubes of chicken liver). I don't hover over my mother's plate with a scale and a calculator, but if she's like an average Pole, she eats about 156 pounds of meat per year. Americans devour even more: 275 pounds a year, give or take. In the meantime, many scientific journals report on the detrimental health effects of eating meat. According to studies, high consumers of cured meats and red meat are at a 20 to 30 percent increased risk of colorectal cancer. High intake of red meat and processed poultry may raise the risk of diabetes in men by 43 percent and in women by 30 percent. In one widely cited study that followed over 120,000 people, researchers associated higher intake of red meat with an elevated risk of cardiovascular and cancer mortality and estimated that "9.3% of deaths in men and 7.6% in women in these cohorts could be prevented at the end of follow-up if all the individuals consumed fewer than 0.5 servings per day (approximately 42 g/d) of red meat." Meanwhile, studies show that the vegetarian Seventh-day Adventists in California live on average 9.5 (men) and 6.1 (women) years longer than other Californians.

Do reports like these deter us from eating meat? Not really. American meat consumption has been growing for decades. According to the US Department of Agriculture (USDA), in 2011 we ate an average of sixty-one pounds more of meat than we did in 1951—that's about 122 average eight-ounce steaks a year *more*, despite all the accumulating warnings about cancer, diabetes, and heart disease and despite the fact that the first of these warnings came as early as the 1960s. And it's not just the United States. Across the world, the appetite for animal protein is on the rise. The Organisation for Ecomomic Co-operation and Development (OECD) estimates that by 2020 the demand for meat in North America will increase by 8 percent (as compared to 2011), in Europe by 7 percent, and in Asia by a whopping 56 percent. In China, meat consumption has quadrupled since 1980. Studies on the deteriorating health of the Chinese caused by their growing meat consumption (among other causes) are mushrooming in scientific journals. But the black scenarios painted by scientists don't seem to scare the Asians away from Kung Pao chicken and Mushu pork.

This international love of animal protein is not only messing up our health, it's also damaging the planet. The media have reported on this over and over: each burger contributes as much to global warming as driving an average American car for 320 miles. Producing one calorie from animal protein releases eleven times more carbon dioxide than producing one calorie from plants. Meat eating is responsible for up to 22 percent of all greenhouse gases—by comparison aviation contributes a mere 2 percent. That's a huge deal. According to some new estimates, global warming may eventually cause sea levels to rise as much as sixteen to twenty-nine feet, flooding cities like New York and Shanghai by the end of this century. And so scientists and politicians (at least some of them) are trying to come up with solutions, thinking up new energy sources, deliberating how to encourage people to consume less, drive smaller cars, and so on. But there is one thing that, in theory, is very easy to do—much easier than, say, inventing solar-powered cars—and that would greatly reduce carbon emissions, slow global warming, and improve our chances of survival. That thing is to go vegetarian. And yet, we don't want to give up meat, New York City be damned.

The meat puzzle has a moral dimension as well. According to a 2003 Gallup poll, 25 percent of Americans claim that animals deserve "the exact same rights as people to be free from harm and exploitation." In one study, 81 percent of Ohioans said that the well-being of farm animals is just as important to them as the well-being of pets. Yet we do not spoil farm animals the way we spoil our pooches and kitties, nor do we guarantee them the same rights as we guarantee humans. Instead, we snip the beaks of caged chickens, without anesthesia, to prevent them from killing each other out of desperation. We cut the tails of pigs short (also without anesthesia) so that they don't bite them off as they lose their minds. We crowd our egg-laying hens eleven to a cage, packed so tightly that they cannot move. As a result they sometimes get stuck between the bars and die of hunger and thirst. It's not that we don't feel empathy toward farm animals or like seeing them suffer. On some level, it does disturb us, and that's precisely why we engage in elaborate mental exercises to avoid feeling guilty over all the harm these cows, pigs, and chickens are fated to suffer. We convince ourselves these animals are less smart than they really are. We disconnect the living creature from the food on our plates. Scientists call it "cognitive dissonance reducing techniques" and show that even attaching the label "meat" to a species means that we start treating it differently, with less respect.

The harm to our health, our planet, and our conscience notwithstanding, the human race is no closer to letting go of meat. According to Gallup, in 1943 the number of Americans who didn't eat meat was about 2 percent. By 2012 the number of people who consider themselves vegetarian had risen to 5 percent (again according to Gallup). But another survey showed that 60 percent of the Americans who are self-described "vegetarians" actually consume red meat, poultry, or fish at least occasionally, which roughly brings us back to 2.4 percent of committed vegetarians—about the same as in 1943.

Myself, I'm one of the sloppy vegetarians. First of all, I eat fish. I do it mostly because I'm lazy. I live in France, a country of *foie gras* and horse steaks. And I'm not talking Paris. I'm talking small village in the middle of a vast forest—very vegetarian unfriendly. I enjoy dining with friends in restaurants, and if I were to stick to meat-free dishes, by now I would have consumed about half a thousand goat cheese salads. There is not much else on local menus that doesn't contain animal flesh. And so I order *poisson blanc au beurre à l'ail* (white fish in butter and garlic sauce) or *saumon aux herbes* (salmon with herbs). But it's not just the fish eating that I'm guilty of. Sometimes, if no one can see me—and this is really difficult to admit—I nibble on a slice of sausage or a strip of bacon. It doesn't happen often—maybe once every six months or so. The taste usually disappoints me. I feel guilty over harming the poor cow, pig, or chicken and swear I'll never do it again. And then, sure enough, I do it again. Just like my mother, I can't seem to completely let go of meat either. There is something in it—in its cultural, historic, and social appeal, or maybe in its chemical composition—that keeps luring me back.

There are many books on the shelves of American bookstores dealing with the unhealthiness of our addiction to meat and at least as many about the suffering of farm animals. I've read most of them, yet none answered the question that kept bothering me: Why do we eat meat *at all*? I wrote this book because I wanted to find out what meat offers people, such that despite its costs—the guilt, the damaged arteries, the polluted planet—we carry on eating it. It seems as if nature has played a trick on us and given us a craving for something that is basically bad for our well-being.

So, what drives us to do it? My mother's answer—"because I like it"—isn't enough. It makes me think of a teenage girl who is dating an inappropriate boyfriend and tells her anxious parents she refuses to leave him because she "loves him." At first glance, it seems like a good response. But she doesn't love the boy "just because." She loves this particular human male because his body gives off pheromones that attract her, because culturally she is predisposed to be drawn toward tall and muscular types, because she was raised by, say, a controlling mother and an insecure father, so she likes her boyfriends to be free-spirited. Likewise, we don't eat meat "just because we like it." There is much more to our meat hunger than that.

This book is an investigation into why humans love eating meat. The story it tells begins 1.5 billion years ago in the temperate waters of Earth's only ocean, when ancient bacteria got hooked on the "flesh" of others. Spanning millennia, it uncovers our planet's first carnivores and their victims, the first-ever meat animals. It follows our hominin ancestors as they learned to eat meat and tracks down the benefits that they derived from becoming part-time carnivores: among them were a larger brain and advanced social structures. Some scientists would go so far as to say that meat eating has actually made us human. Not only did it help us migrate out of Africa but it was even behind our thinned hair and profuse sweating (relative to our cousins, the chimps).

As we approach the modern era, this book turns to biochemistry. Is there something in meat's chemical composition that keeps us hooked? Is it the 2-methyl-3-furanthiol or one of the other one thousand volatile compounds that together make up the specific, mouthwatering scent of cooked meat? Is it the *umami* taste (Japanese for "delicious") that is found mostly in meat, mushrooms, and milk? Or is meat actually necessary for staying healthy? Despite the risks of cancer and heart disease, what if the human race would be even worse off without meat, a planet full of small, immune-deficient weaklings? Are some people, those with a gene mutation that makes them dislike the scent of androstenone (a mammalian pheromone), destined to be vegetarians, while others, those who are particularly sensitive to bitter compounds in fruits and vegetables, more likely to love meat? Is it the skillful marketing and lobbying of the powerful meat industry, with its $186 billion worth of annual sales in the US alone, that keep us hooked on animal protein against our best interests? Or maybe, just maybe, do we eat meat simply out of habit, because it got so engrained in our culture and history that we just cannot let go of it? After all, what would Thanksgiving look like without a turkey or a summer grill without a burger? Do we eat meat because over the centuries it has come to symbolize masculinity, power over the poor, power over nature, and power over other nations? Is our love of meat a kind of addiction—psychological, chemical, or maybe a little of both? And if it is, will we ever be able to break it? Is telling people to "cut down on meat" no different from telling a chain-smoker to go cold turkey?

As this book reveals, there are many reasons why meat is so attractive to us. I call these reasons "hooks." The hooks are linked to our genes, culture, history, the power of the meat industry, and the policies of our governments. I examine these hooks in detail, one by one, to discover the individual reasons for meat's appeal—such as the importance of a particular polymorphism of serotonin receptor genes 5-HT that can affect how much beef you eat, or the role $2.7 billion in corn subsidies plays in boosting American appetites for meat. In each chapter of the book, I analyze the hooks, big and small. I conclude by showing the likely future of humanity's relationship with meat: Are we ever going to cut down our meat consumption? What happens if we don't? Are we going to soon be eating lab-grown steak chips, insect burgers, or plant-based chicken that we 3-D print in our own kitchens?

Meathooked is not a book about the detrimental health effects of meat consumption, nor is it an essay on the suffering of farm animals. There are enough of those already. I may be a vegetarian, but I won't tell you how much meat you should or shouldn't eat. I'll just give you the facts: what's in the taste of meat that keeps us hooked, how our culture encourages meat eating, how deeply the need to consume animals is engraved in our genes. The rest is for you to decide.

If you are an avid meat lover, this book will help you understand what drives your appetite and will make you aware of the ways meat eating influences who you are and how you behave. If you are one of the 39 percent of Americans who is trying to cut down on meat, this book can help you change your diet and show you the reasons why reducing meat consumption may be difficult and what you can do about it. Just as it may be hard to stop smoking if you don't know why you got addicted in the first place, it may be hard to give up meat if you don't know *why* you crave it. And for committed vegetarians and vegans, this book offers an insight into why the majority of humans don't follow in your steps and so often react with anger if encouraged to do so. I've written this book hoping it will help you make conscious, informed decisions about your diet, instead of simply adhering to the eating scripts written for us by culture, habits, imperfect government dietary guidelines, or what your mother ate during pregnancy.

But above all, this book will be a story—a story that I'm hoping will entertain you as it takes you through history and through space, from the depths of the Precambrian to the mid-twenty-first century, from steak houses in India and voodoo temples in Benin to the meat labs of Pennsylvania. It will be a story about humanity's love affair with meat: how it started, why it continues going on so strong, and how it may end—if it ever will.

I

Enter Meat Eaters

In a way, the history of life on Earth is a history of eating meat. It's a story of cheating, of growing larger and larger, of trying to hide. It's a story of an arms race between predators and their prey. This story begins about 1.5 billion years ago, in the temperate waters of Earth's only ocean. There were no animals back then, no creatures with complicated body plans—no legs to walk on, no hearts to pump blood, no teeth to shear off meat. And there was no meat, either, not in the common sense of the word—the edible flesh of animals was still a long way off.

All life on Earth was simple and single celled 1.5 billion years ago. Only two types of organisms probably existed: bacteria and archaea. The latter are bacteria-like creatures that are nowadays known for their ability to live in extreme environments, such as in deep-sea thermal vents where temperatures reach over 212 degrees Fahrenheit, in the supersaline waters of the Dead Sea, and even in petroleum deposits. For these ancient bacteria and archaea, the world might have resembled a garden of Eden, with no predators and no killing. They fed themselves on energy from the sun or from inorganic sources such as elemental sulphur or hydrogen. But the peace was soon about to end.

According to Gáspár Jékely, a youngish researcher at the Max Planck Institute for Developmental Biology in Germany, the story of predators and carnivores on Earth began with cheating. Ancient bacteria were unable to eat one another, Jékely says. Bacteria don't have mouths. To consume something, a bacterium has to engulf its prey with its entire single-cell body in an act known as phagocytosis. But the problem is that the cell of a bacterium has a rigid wall, a bit like bark on the trunk of a tree, which prevents it from opening up and swallowing the flesh of others. Bacteria can't easily get rid of the wall, either. If they did, they would expose themselves to the forces of the outside world, which could mean death. And yet, at some point around 1.5 billion years ago, some of the ancient bacteria did begin to shed their cell walls and did start eating others. They could do it because they were cheaters.

As Jékely told me, bacteria are social creatures. In the Garden of Eden, they lived communally, just as they often do nowadays, in foam-like, slimy biofilms that floated on water or covered the rocks on the bottom of the ocean. In such biofilms each bacterium has to secrete something for the common good. "It's like a community building a house, where everybody is supposed to bring a brick," Jékely said. But some cheat. "They just pretend to bring the brick, and yet still live in the house. They exploit the community." In the safety of the biofilm, such cheating bacteria were able to get rid of their cell walls and become predators. Instead of synthesizing energy from the sun or from inorganic compounds, they would crawl up, amoeba-style, to other bacteria and swallow them whole. It made perfect sense. Eating others was an efficient way of getting nutrients.

Of course, phagocytosis done by ancient bacteria was hardly like modern meat eating. These, though, were the first *predators*, organisms eating other organisms by killing them, and even some scientists call these earliest predators "carnivores," as if what they were doing was indeed devouring meat.

These first acts of predation, many researchers agree, held great consequences for life on Earth. They were essential to the emergence of eukaryotes, organisms with complex cells containing organelles. After the predators started hunting down other bacteria, the engulfed prey would sometimes develop defense mechanisms to avoid being digested and would survive inside the predator. With time and generations, these fortified prey would evolve into organelles such as mitochondria—and eukaryotes emerged. All animals and all plants are eukaryotes.

Once the ancient bacteria got hooked on the "meat" of others, it started a chain of events that led not only to the emergence of eukaryotes with their complex cells but also to many other vital transitions in evolution: going from unicellular to multicellular (it's harder to get eaten if you have multiple cells instead of just one), from small to large, from soft bodied to hard shelled, from slow to fast. Without that first bacterium eating up another, there would have been no eukaryotes on Earth, no multicellular organisms, no animals, no meat eaters, and no meat.

The game of life was changing. Soon enough, specialized predators would be hunting the muscled flesh of others, and meat eating as we know it today would be born.

In the warm oceans of late Precambrian Earth, some time around 550 million years ago, one of the first true carnivores began eating meat. We know of this carnivore's existence because it left traces in the fossilized carcasses of animals called *Cloudina*. But this ancient predation couldn't be further from the primal scene we might imagine of a shark-like predator in hot pursuit of a dolphin-like victim.

Although we still don't know the identity of this Precambrian carnivore, it almost certainly wouldn't have done much damage to a modern human. It was probably rather small (about 0.02 inch long—like a short grain of rice), and instead of ripping its prey apart with a mouthful of sharp teeth, it bored holes into it. The prey wasn't exactly fleeing in terror, either. *Cloudina* was an anemone—a coral-like animal—and formed shells that resembled towers made of shot glasses. It lived out its life attached to the bottom of the sea, which is where it met its fate, eaten by a creature that drilled holes as thin as an average human hair into its shells.

This may not sound much like carnivorous behavior. Still, if we define a carnivore (after *Encyclopedia Britannica*) as an "animal whose diet consists of other animals," then the predator of *Cloudina* most likely fits the bill. But was there even any actual meat to be eaten on a *Cloudina*? After all, an anemone doesn't look much like a sirloin steak. Although the word *meat* is usually understood to mean the edible parts of an animal, the most important component of meat is skeletal muscle—muscles that we can contract voluntarily, as opposed to heart muscle and the smooth muscles that make up blood vessels, the bladder, or the uterus. Did *Cloudina* have skeletal muscles, a real meat to be eaten? Quite likely, yes. Scientists believe that skeletal muscle has been around for at least six hundred million years. Jellyfish-like animals, of which *Cloudina* was most likely one, were the first to evolve skeletal muscles—and so they were the first meat animals on Earth. Of course, how that meat would have tasted remains a mystery, but it would have probably been a bit similar to modern anemones, which are eaten in Chinese and Spanish cuisines and which, according to one present-day food blogger, taste like "a hybrid of pork and veggies with a fishy aftertaste."

Odds are that the mysterious, hole-drilling predator of *Cloudina* wasn't even the first meat eater on Earth. But it was the first meat eater whose traces we have found so far, even though we still have no idea what it might have looked like.

The first carnivores that we can actually identify appear later. One of those earliest meat eaters still exists today and is known as the penis worm. Disturbing as the name may seem, if you search the web for images of these animals, it becomes obvious they were named for their looks: they were long, sausage-like, pale pink, and considerably thicker on one end. Yet penis worms weren't fierce, Jurassic-like predators, either (an image that would have been even more disturbing). They resembled tubes through which food passed, and they fed on almost anything in their path, including shrimp-like arthropods, cone-shaped hyolithids, and trilobites—basically, meat, in all of its early forms.

With the appearance of another strange new carnivore, the story of meat eating becomes a bit more gripping than drilling holes in shells or sifting through ocean sediments. The squid-like *Nectocaris*, although as bizarre in its looks as the penis worm, was probably a far more skilled carnivore. It had two tentacles, effective for manipulating prey, a conveyor-belt-like tongue with teeth on the surface, eyes on long stalks, and a weird-looking funnel, used to squirt itself around. No longer than six inches, *Nectocaris* may be a rather small predator by today's standards, but it was quite big for the early Cambrian. What did it hunt? *Nectocaris* squirted itself after little shrimp-like animals, mollusks, worms, and maybe a jellyfish here and there. Fierce? Dangerous? If you are a small Cambrian mollusk, then certainly yes.

As time passed, meat-eating predators got bigger. By the mid-Cambrian, about five hundred million years ago, *Anomalocaris* entered the scene. It was truly large and fierce: three feet long, its body was streamlined for fast movement, complete with stalked eyes for clear vision and a round mouth full of sharp, teeth-like plates. It was the largest meat eater of the Cambrian and the first known apex predator, a carnivore that sits at the top of its food chain. In its time, *Anomalocaris* was the king of meat eaters.

An evolutionary arms race had begun. Once predators like *Anomalocaris* and *Nectocaris* (and yes, the penis worms, too) got hooked on meat, the battle between predator and prey became one of the driving forces for evolution, leading to the Cambrian period's explosion of biodiversity.

It worked like this: a big, hard-bodied animal is better off than a small, soft-bodied animal, which is more easily snapped up by a passing meat eater. It's a good idea for prey to grow larger so that a predator won't be able to swallow it. On top of that, a nice hard shell offers protection, too. Once prey animals hid themselves in shells, the predators had to find ways to get to them anyway. They drilled holes; they grew teeth-like plates and sharp, conveyor-belt tongues. As their prey increased in size, so did the predators. First a few inch-long penis worms appeared, then six-inch *Nectocaris*, then over three-foot-long *Anomalocaris*, and, down the evolutionary road, the enormous meat-eating dinosaurs, some of which were as long as eight male lions put in a row. The whole animal kingdom started to bulk up, to invent new ways of eating and of avoiding being eaten.

There are other factors behind the world's growing appetite for meat and the proliferation of species that followed. Some scientists believe that the boom of carnivory that started in the Cambrian couldn't have happened if the oxygen levels on Earth, and especially in Earth's oceans, remained low. In the period preceding the Cambrian, the oxygen levels in the atmosphere were only about 15 percent of what we have now, which means that if you time-traveled back to 650 million years ago, you would have suffocated within minutes. To truly blossom, carnivores need oxygen. Chasing prey is energetically costly, as is digesting big chunks of meat. Even today, there are relatively few meat eaters in the oxygen-poor waters of the oceans. According to one hypothesis, once the climate got warmer and the glaciers that had enveloped Earth before 650 million years ago started to melt, large amounts of nutrients were released into the oceans, which increased the population of tiny algae, which in turn produced more oxygen. This would have given the meat eaters the necessary boost to proliferate, speeding up the arms race. If Earth hadn't become well endowed with oxygen, it seems, it wouldn't have become the planet of meat eaters it is now.

The next chapter of the story of carnivory on Earth—and the tale of how humans became such avid meat eaters—begins sixty-five million years ago. The dinosaurs have just gone extinct, together with over half of Earth's species. In rain forests that carpet vast areas of the planet, among soaring trees ribboned with vines, the next line of our ancestors has just evolved. It's the first primate ever known—*Purgatorius*. It doesn't look much like you or me, or even like a chimp. It resembles a cross between a mouse and a squirrel. And if it were still alive today, it would likely pass for a cute pet.

Purgatorius was an accomplished tree climber—and a vegan. It gave up the insect-based diet of its ancestors in favor of newly abundant fruits and flowers, carving for itself a comfortable niche high in the branches. For tens of millions of years, the descendants of *Purgatorius*, some of which would later evolve into us, were committed to their plant-based diets. From small monkeys to gorilla-size apes, they survived mostly on tropical fruits—spicing their meals with occasional worms, often by accident. About fifteen million years ago, they diversified a bit, adding hard seeds and nuts to their diets, but stayed true to their vegan roots.

Then, around six million years ago, *Sahelanthropus tchadensis* entered the African primate scene. With the advent of *Sahelanthropus*, our lineage likely split from that of our closest cousins, the chimps and bonobos. In the language of paleoanthropology, the word *hominin* stands for modern humans and all the extinct species closely related to us—and *Sahelanthropus* was the first. A short, flat-faced, small-brained creature, it most likely walked upright on two legs. It had smaller canine teeth than its ancestors and thicker tooth enamel, which suggests that its diet required more chewing and grinding than *Purgatorius*-like meals of fruits and flowers.

Nevertheless, meat eating still hadn't caught on among our ancestors. *Sahelanthropus* probably ate tough, fibrous plants supplemented with seeds and nuts. The several species of *Australopithecus* that lived between four and three million years ago in woodlands, riverine forests, and on seasonal floodplains of Africa weren't hooked on meat, either. Their dental microwear—the pattern of microscopic pits and scratches left on the surface of their teeth by the foods they ate—suggests a diet similar to that of modern chimps: some leaves and shoots, lots of fruits, flowers, a few insects here and there, and even tree bark. Did australopiths ever eat meat? It's possible. Just as modern chimps occasionally hunt colobus monkeys, our ancestors may have occasionally dined on the raw meat of small monkeys, too. Yet the guts of early hominins wouldn't have allowed them to have a meat-heavy diet, like the one Americans eat today. Their guts were characteristic of fruit-and-leaf eaters, with a big caecum—a bacteria-brimming pouch at the beginning of the large intestine. If an australopith gorged himself on meat—say, ate a few zebra steaks tartare in one sitting—he likely would have suffered twisting of the colon, with piercing stomach pains, nausea, and bloating, possibly resulting in death. And yet in spite of these dangers, by 2.5 million years ago, our ancestors had become meat eaters.

It seems that our bodies had to adjust gradually—first getting hooked on seeds and nuts. Seeds and nuts are rich in fats but poor in fiber. If our ancestors ate a lot of them, such a diet would have encouraged the growth of the small intestine (where the digestion of lipids takes place) and the shrinking of the caecum (where fibers are digested). This would have made our guts better for processing meat. What's more, a diet of seeds and nuts could have prepared our ancestors for a carnivorous lifestyle in another way, too: it could have given them the tools for carving carcasses. Some researchers suggest that the simple stone tools used for pounding seeds and nuts could have easily been reassigned to cracking animal bones and cutting off chunks of flesh. And so, by 2.5 million years ago, our ancestors were ready for meat: they had the tools to get it and the bodies to digest it. But being capable is one thing; having the will and skill to go out and get meat is quite another.

So why would our ancestors have made such a shift? What inspired them to look at antelopes and hippos as potential dinners? The answer, or at least a part of it, may lie in a change of climate approximately 2.5 million years ago. As the rains became less abundant, so did the fruits, leaves, and flowers that our ancestors relied on. Much of the rain forest turned into sparsely wooded grasslands with few high-quality plants to eat but with more and more grazing animals. During the long, dry spell from January through April, our ancestors would have had problems getting enough food, and to find their usual fare, they would have had to expend more time and calories. Early hominins were at an evolutionary crossroads. Some, like the so-called robust australopiths, chose to eat large quantities of lower-quality plants; others, like early *Homo*, went for meat. The robust australopiths ended up extinct, but early *Homo* survived to evolve into modern humans.

Interestingly, while the ancestors of humans chose to profit from the new wealth of savanna herbivores and their flesh, the ancestors of chimps and gorillas never did. One of the reasons might have been their inability to walk on two legs. Searching for meat is costly, requiring more long-distance walking—and, in turn, more energy—than eating grass or fruit. Moving on two legs is more energy efficient than chimp- or gorilla-style knuckle walking, and longer legs better dissipate temperature, which prevents overheating and boosts endurance. It seems that if *Sahelanthropus* or its ancestors didn't stand up straight (or at least straight-ish) six million years ago, a few million years down the road early *Homo* wouldn't have been so well equipped to search for meat and might not have developed a taste for animal flesh—and there might not now be steaks or burgers on the dinner tables of today.

Still unanswered, however, is the question of what actually happened: Why was it that one day our ancestors were passing the animals grazing the savanna without a second thought, and the next day they saw them as food? Maybe a few of our ancestors were walking among acacia trees and saw a saber-toothed cat feed on a gazelle. Maybe they stumbled upon a dead zebra, with its guts spilling out and meat exposed, and thought, hey, why not give it a try? Even dedicated herbivores such as deer or cows will sometimes try meat if they chance upon it. There are records of cows devouring live chicks and munching dead rabbits, of deer eating birds, and of the duiker, a tiny African antelope, hunting frogs. (If you want to see a few of these carnivorous herbivores caught on camera, check out YouTube.) So it comes hardly as a surprise that our ancestors, who might have already been supplementing their diets with the meat of an occasional small monkey, saw the new abundance of savanna grazers as a way to get a few additional calories. The hominins were already omnivorous and opportunistic. If something was edible and it was there, they ate it. By 2.6 million years ago, there was a lot of meat around. Just as *Purgatorius* took advantage of the climate change and a new wealth of fruits, their descendants, early *Homo*, successfully adapted their diets to the changes in their environment. But this time it meant going after meat.

There are bones all around me: elephant bones, jaws of saber-toothed cats, a few skulls of extinct hyenas, even hominin skulls. I'm standing in the lab of Briana Pobiner at the Smithsonian National Museum of Natural History in Washington, DC. Pobiner studies bones, which she either digs out of archaeological sites in Africa or steals from lions, to understand how and when our ancestors started to eat meat.

As we talk, Pobiner opens a drawer and pulls out a rib of an elephant, a million years old and scarred with cut marks by one of our stone-tool-wielding ancestors. Tracing her finger along a groove, she explains that human-made cut marks differ from the marks left by the teeth of a lion or by water dragging the bones across rocks. "Cut marks are V shaped, more linear in comparison to carnivore tooth marks and deeper than sedimentary abrasions," she says, and then shrugs. "This one is pretty obvious, but sometimes it's rather hard to tell."

For modern scientists, cut marks are important because they are the earliest hard evidence of our meat eating. Yet for our ancestors, cut marks were just mistakes. Instead of smoothly slicing meat, some prehistoric butcher cut the bone with his stone tool, leaving a mark. Today, by studying these cut marks, researchers like Pobiner can tell what our ancestors ate, whether they hunted or scavenged, which parts of an animal they usually consumed, and whether the butchers were skilled professionals or amateurs. It's a whole story written in a Braille-like language.

The oldest undisputed record of cut marks tells us that humans started to butcher savanna animals 2.6 to 2.5 million years ago. Someone back then in what is now Ethiopia filleted a *Hipparion*, an extinct three-toed species of horse, and cut out the tongue of a medium-sized antelope. We have no idea, though, if this was just an occasional foray into carnivory, a once-in-a-lifetime event, or if the butchers were eating meat on a regular basis. But by two million years ago, meat appears to have entered the diets of our ancestors for good. Pobiner and her colleagues have recently found evidence of what she calls "persistent carnivory" in Kenya. "These early humans came back to the same place, over and over, to butcher and eat animals," she tells me.

Our ancestors weren't particularly picky eaters. They ate their way through the kingdom of savanna grazers: they butchered warthogs, small gazelles, rhinos, giraffes, waterbucks, elephants, and a few extinct species, too. One such prey, *Hippopotamus gorgops*, was a larger cousin of living hippos, with rather bizarre-looking eyes perched on long stalks. Another, *Deinotherium* (Greek, meaning "terrible beast"), resembled a cross between a giant elephant and an anteater. Some of the butchered animals were truly impressive in size, weighing up to 5,500 pounds. Some were much, much smaller (think hedgehogs). In the minds of our ancestors, almost any animal was meat—even fellow hominins: a few hominin bones scarred with cut marks have been identified as proof of cannibalism.

The big question scientists continue to argue about is how much of the meat that our ancestors ate about 1.8 to 1.5 million years ago was scavenged and how much of it was actually hunted. To find some answers, scientists study not only the pattern of the cut marks but also the behavior of modern carnivores. In Pobiner's case, this means driving around the East African savanna in a Toyota Land Cruiser with an armed guard, seeking out recent kills made by lions and leopards. Once a predator is done eating, Pobiner hauls the leftover, bloody carcass of a zebra or an antelope into the trunk of her car (she has to take out the backseats to make room for the dead animals). Later, back at the camp, her assistant boils the carcass to clean the bones so that Pobiner can study the tooth marks and the damage that the carnivores have inflicted. Pobiner also weighs the meat to calculate how much is left over. "The lions leave *lots* of meat," she tells me, stretching the *o*'s in *lots*. That's an important finding because it means a feast of animal flesh didn't necessarily require our ancestors to go hunting—stealing the kills of big carnivores would have worked just fine.

And at the start of our ancestors' serious meat eating, that's probably what they did. They could have found a carcass of a giraffe abandoned by a lion but not yet discovered by hyenas or vultures or, more likely, a carcass of an antelope pulled up a tree by a leopard (leopards hide their untouched kills in trees for a later meal). Since our ancestors could still climb pretty well at this point, they had an obvious advantage over other scavengers. On the ground, they would have to compete for meat with a multitude of other hungry creatures: jackals, hunting dogs, lions, hyenas. But even if they weren't lucky enough to arrive before other scavengers did, there was still a good chance for leftover brains and bone marrow. Although brains and bone marrow may not sound particularly appetizing to modern humans, Westerners in particular, our ancestors would have considered them a lucky find because they are fatty and loaded with calories. Marrow from a tiny, thirty-pound gazelle would have provided about five hundred calories—as much as a large serving of french fries at McDonald's. A scavenged wildebeest would have meant six times that.

Although we might have started as regular scavengers (what scientists call "passive" scavengers), who grabbed whatever flesh they stumbled upon, most likely quite early on hominins became "power" or "confrontational" scavengers and stole the kills right from underneath the noses of hungry lions, leopards, and saber-toothed cats. The pattern of cut marks on some of the ancient butchered bones suggests that once in a while early *Homo* ate those parts of prey that carnivores prefer as their first bites. The cats had little chance to munch on whatever they had just hunted, because coordinated groups of our ancestors chased them off and won the dinner. The hominins would then drag the meat to their day camp for butchering and sharing. They would start with the parts they liked best: fleshy limbs and fatty tongues. But to do the butchering, they needed tools: sharp flakes to scrape the meat off the bones, bigger stones to hammer them open to extract the marrow. Without stone tools, our ancestors couldn't have become full-time meat eaters. They just didn't have the bodies of carnivores.

One claim in particular gets repeated over and over on forums and blogs: we have pointy canine teeth, which means we are "designed" to eat animal flesh. Yet that is simply not true. Yes, we do have canine teeth, but that's not proof that we are made for meat-based diets. Canine teeth are just one of the basic kinds of teeth of mammals. Most mammals have them, including such plant chewers as deer and horses. The water deer, native to China and Korea, have large, sharp canines, over two inches long, and so formidable that they look like they might belong to a saber-toothed cat.

Besides, human canines are not exactly sharp and pointy, as some bloggers and forum writers claim. In fact, they are relatively small and stubby. Even though canine teeth in apes (including humans) are indeed used for slicing and shearing food, their size and shape don't have much to do with diet: they have much more to do with sex and fighting. Take gorillas. They eat primarily leaves and fruits but have dagger-like canines—especially the males. They don't need these canines to tear apart flesh; they need them as weapons against other gorillas, particularly when battling over who gets the females. Fighting with members of their own species is also why the water deer have such weirdly big canines. Most likely the canine teeth of our ancestors shrank in size (which allowed space for bigger molars—better for chewing tough plants) because they didn't fight as much among themselves as other apes did and because they were much more monogamous. Also, development of weapons enabled our ancestors to forgo big canines. They no longer needed to bite each other; they could pierce each other with spears instead. This gradual scaling down of hominin canines has been happening for at least six million years. Our early ancestor *Sahelanthropus* already had reduced canine teeth. These smaller canines are actually considered to be one of the main features distinguishing our ancestors from other apes. Rather than proof of some inherent carnivorous nature, our canine teeth are a sign that, for better or for worse, we should stick with the same mate.

The true meat-eating teeth are not the canines but the carnassials. If those sound unfamiliar, it's because we humans don't have them. Cats, dogs, and even skunks have carnassials. If you manage to open the muzzle of a friendly Fido, you will see them at the back of the jaw: they are blade-like and sharp and perfect for slicing meat. Carnassials are the key characteristic of members of the order Carnivora, which includes primarily predatory mammals, such as lions, tigers, seals, raccoons, and domestic cats.

What we humans also don't have are the jaws of carnivores. Watch any savanna-based nature documentary, and you'll likely see a lion roar. These cats can really open their jaws wide. Hominin yawns, both those of modern humans and our ancestors, don't even come close. We also lack the impressive temporalis muscles of lions (that's the muscle that will hurt if you chew gum too long). All this means that we not only can't kill prey with our mouths, carnivore-style, but we also have trouble eating raw, unprocessed meat.

If you are a hominin, whether an early *Homo* or a modern human, and you find a dead, untouched zebra on a savanna, unless you have some sharp tools with you, you have a problem. All that zebra meat is hidden inside a thick wrapping—skin—that your teeth are completely unable to break. Just imagine taking a bite of a living cow. Although our ancestors had bigger teeth than we do, their teeth weren't much better when it came to chewing through fur and skin. You can rip apart a small monkey with your hands, chimp-style, but to get to the meat of a dead giraffe, you would need the help of true carnivores. You would need to wait, as vultures do, until other, better-equipped animals tore through the skin of your dead zebra, exposing the meat. Or you could simply wait a bit longer. With time, the carcass would rot, and the skin would become easier to break. But even if you did manage to get a chunk of zebra in your mouth, your blunt teeth would likely be unable to reduce the meat into pieces small enough to swallow. Before the invention of stone tools circa 2.6 million years ago, early humans were simply unable to hunt big game. Their bodies were just not equipped for that.

But thanks to tools, our ancestors likely became quite skilled at hunting small gazelles as recently as 1.8 million years ago. Although scientists have yet to find the weapons used to hunt them, the mere presence of the bones of such animals (bearing cut marks suggesting they had tons of meat on them) likely means that they were hunted, not scavenged. Since lions and hyenas can devour a tiny gazelle in minutes, hominins would not have had enough time to steal the kill.

How did we hunt? If the prey was truly small, we could have done it with our bare hands. The famed British-Kenyan paleoanthropologist Louis Leakey, who in the early twentieth century excavated Tanzania's Olduvai Gorge, was proud of his ability to catch small hares and antelopes by hand and reasoned that if he could do it, our ancestors probably could, too. Another possible scenario involves acacia trees. Acacias are widespread on the African savanna and have branches spikier than barbed wire. Their thorns are the reason why locals surround their livestock with acacia branches for nightly protection against predators. They are also the reason why one-eyed giraffes are not uncommon: they feed on acacia leaves, and accidents happen. At first, our ancestors could have used acacia branches as a power-scavenging tool to scare away lions or cheetahs from their kills. From there, it's not a long way from waving a thorny branch in front of a carnivore's nose to stabbing an antelope with a branch that has been sharpened at one end. Once early *Homo* had stone tools, they could have used them to make the tips of their acacia branches more pointy and deadly. You don't need a particularly big brain to come up with such an idea. Chimps did, after all. In the jungles of Senegal, groups of chimpanzees use their teeth to sharpen sticks into spears, which they then use to stab bush babies—tiny primates as cute as their name suggests.

Once hominins controlled fire, they could have used it to further harden the tips of their spears. Shoving a stick into a campfire makes it much more resistant—and much more lethal. But it probably took our ancestors quite a long time before they invented the stone tips that would make their spears even more efficient. The earliest evidence of such weapons comes from a site in South Africa, five hundred thousand years old. Thus it seems that we hunted for over a million years with little more than sharp sticks. This seems almost impossible, considering we are not particularly big animals (and our ancestors were even smaller), we don't have claws, and we're not very fast. So how were we able to kill prey in the years before stone-tipped spears? The first reason, and the most obvious, was that we were social creatures who hunted cooperatively. The second was our ability to climb trees.

Imagine a warm, sunny day on an East African savanna a million and a half years ago. But don't think of the Serengeti—imagine a landscape much more wooded. Numerous trees and shrubs dot the landscape, providing shade, food, and places to hide. Among the trees, a gazelle is slowly munching on the grass, unaware that high among the branches a group of hominins is waiting, preparing to strike. The gazelle moves closer and closer. Suddenly, the apes cast their spears. A few of the pointy sticks pierce the prey's body. The animal falls, and gazelle is on the menu.

With time, hunting became part of life for early humans. We "invented" sticks, then discovered how to sharpen them and harden them in fire and how to add stone tips. A spear found between the ribs of an elephant in Germany dating to about 400,000 or 300,000 years ago (yes, there were elephants in Germany back then) proves that by that time our ancestors were skilled crafters of javelins—light spears designed for efficient throwing. Over 7.8 feet long and made of the hardest part of yew wood, from near the base of the tree, they were shaped so that the maximum thickness and weight were about a third of the way from the tip, which allowed for the easiest throwing. Our ancestors had learned the art of killing large animals at a distance.

Yet, as many researchers point out, we should not overromanticize hunting. Modern cultural mores cause us to look at hunting as something noble and evolved, while shrugging off scavenging as dirty, easy, and almost indecent. So it is that we tend to think of lions as the kings of the jungle—they are the hunters after all—and of scavenging hyenas as lowly and cowardly. But there is no reason to assume that scavenging is somehow a lesser way to obtain meat. Lions scavenge frequently—in the Serengeti more than 40 percent of their diet is stolen from other hunters. And hyenas hunt more often than they scavenge. Scavenging is both difficult and dangerous. To be a good scavenger, you need to watch out for signs of a kill, you need to arrive at the site before your competitors do, and you may even need to fight off the hunter or other scavengers. Thinking of hunting as a superior, more virtuous way of obtaining meat is just our present-day bias.

Even if early on we were "only" scavengers, 2.5 million years ago we were definitely already meat eaters. We got hooked on meat because of climate change and because our usual fare became harder to find. Simply put, we went for meat because it was there. Just as it had been for the cheating bacteria in the Garden of Eden, swallowing others was a more efficient way to get nutrients, and scavenging dead antelopes helped our Paleolithic ancestors fend off hunger in times of scarcity.

Meat had entered the hominin diet for good—and this sea change would have immense consequences. Once early *Homo* started to hunt, a chain of events was set in motion that led to profound changes in our bodies, in our societies, and in our lifestyles. From the earliest days, meat was not just about nutrition. It was about politics and sex, too.

2

Big Brains, Small Guts, and the Politics of Meat

If you search the Internet for images of "Paleolithic hunting," your computer likely won't show you early *Homo* chasing hedgehogs. Instead, you will mostly find pictures of mammoths, some elephants, and a few giant rhinos. We seem to have this romantic image of our ancestors taking down only the fiercest and largest of beasts. Does this mean that big game was the best source of nutrition for *Homo erectus* or, later, for *Homo sapiens* and the Neanderthals? Not really. Man's love affair with meat, it appears, was as much about politics and sex as it was about nutrition—if not *mostly* about politics and sex. Take the story of mammoths, for example.

Mammoth Helmut (so named by his discoverers) is packed into dozens of boxes, stacked high in the French National Institute for Preventive Archaeological Research in Paris. I'm meeting Helmut's two young discoverers, Grégory Bayle and Stéphane Péan, in front of silicone replicas of the scattered bones, which are so realistic my dogs would probably try to chew them. Just half an hour earlier, Bayle and Péan announced at a small conference that they had found an incision on one of Helmut's ribs, an incision that might be a cut mark made by Neanderthals scraping off the mammoth's meat. What's so unusual about it (if it's indeed a cut mark) is that we have few such direct proofs of hominins butchering mammoths. That is why many scientists believe that the popular image of our ancestors as expert mammoth hunters is not exactly accurate. There is quite a lot of evidence showing that Pleistocene hunters went mostly for midsized animals, such as bison or reindeer. And yes, sometimes for hedgehogs, too.

Even when humans did hunt mammoths, it might not have always been for their meat. In 2008, far north on the Yana River in Siberia, a massive grave of at least thirty-one mammoths was discovered. The animals were indeed killed by humans—but not necessarily for their flesh. The scientists working on this Siberian site claim that the true booty was the ivory. Just like Greenland Inuits, who used the tusks of narwhals to make spears when wood was scarce, the Siberians of yesteryear used the tusks of mammoths. The meat was just a nice "extra," not the goal in itself.

Péan, Helmut's researcher, believes that there were some Paleolithic cultures related to modern humans in Central and Eastern Europe that did hunt mammoths occasionally. In the spring, as the ground thawed and turned soggy with mud, it became difficult for large animals to plod through, so they were easier to kill. But even these mammoth hunters didn't chase the likes of Helmut solely for their meat. "Hunting mammoths could provide other supplies, such as large bones and tusk ivory to produce varied tools, personal adornment, and art items. It was probably also important for cultural reasons for a group of people to know they can take down such a powerful animal. It might have made them feel stronger and more united," Péan told me.

In most cases, hunting big game such as mammoths was not necessarily the best way to support a family. It's true that one dead elephant is equivalent to about five hundred thousand calories, as much as 909 Big Macs. But hunting hippos, buffaloes, or other large game is highly unpredictable and dangerous. Some anthropologists argue that if hunter-gatherers, whether Paleolithic or modern, truly wanted to provide well for their wives and children, they would do better to go for small animals, even insects, and gather seeds and nuts rather than run after elephants.

Killing large game is difficult. The modern Hadza hunters of Tanzania, with their high-powered bows and poisoned arrows (technology that was unavailable to early *Homo*), fail to bring meat home on 97 out of 100 hunting days. For an hour of work, Hadza men manage to provide on average only 180 calories—that's less than children harvest from gathering. And yet, Hadza men are not the least efficient providers. Hunters from one New Guinea tribe actually expend *more* calories hunting than they manage to get from their kills. They'd be better off just sleeping the whole day in the camp.

Another clue that chasing big game is not exactly about nutrition is the timing of the hunts. If hunting big game was about providing food for your hungry family, logic would dictate that you hunt when there is little else to eat and stomachs are empty. Yet that's often not the case—and probably wasn't in the Paleolithic, either. Modern hunters go after large animals not in times of scarcity but in times of plenty. The San hunters of Botswana, for example, leave on hunting expeditions exactly when tsin beans and mongongo nuts—great sources of calories, proteins, and fat—are the most abundant. Even chimps are more likely to hunt when there are lots of other foods to eat.

What's the point, then? Big game hunting is not really about keeping stomachs full. It's about showing off, politics, and sex. Scoring an elusive prize signals to others that you are strong, skilled, and fearless and that you would make a powerful ally and an adversary to dread. When a hunter *did* bring home an elephant, it was a great source of nutrition, especially of the highly cherished fat. Five hundred thousand calories is a lot of food. There were no freezers back then, of course, and people didn't always know how to preserve meat by smoking or salting it. It was better eaten fast. No one can devour an equivalent of almost a thousand Big Macs in a matter of days, but a tribe can, and so the sharing began.

Many social animals share food. Monkeys do it, ravens do it, bats do it, even whales do it. To reinforce bonds, African hunting dogs eat food vomited by other members of the group, and alpha-male chimps use gifts of meat to show favors and establish coalitions. For early *Homo* meat probably played a similar role. Bringing an elephant to the camp was the Paleolithic equivalent of winning the lottery and writing a big check for charity: it showed that you were a valuable contributor to the public domain, a good neighbor. But hunting big game and sharing the bounty wouldn't carry much social status if the hunters were unable to communicate. How, otherwise, would people back at the camp know who threw the spear that finished the rhino? How would they know which hunter was brave and which stayed back in fear? Chimps share meat only with those who saw the kill. Humans share it with everyone. To build a reputation and play complicated politics, you need to be able to tell others about the hunt. Some scientists believe that big game hunting could have only emerged among those hominins who could talk. Without language, there would have been no mammoth hunts, no prestige linked to meat, no showing off.

And then, there is sex. From the sky, Juruá River in Brazil looks like a twirling, yellowish ribbon crisscrossing a sea of greenness. On its banks, amid the teeming jungle, live the Kulina, a small tribe of people with sharp cheekbones and wide noses. The Kulina practice an unusual ritual they call "order to get meat." Whenever the women of the tribe feel "hungry for meat," they tell the men to go out hunting. At dawn, the women, their straight dark hair hanging loose, stride around the village going from one simple, pillared house to another, rapping at them with sticks and waking the men up. As each man rises from his hammock, the woman who has awakened him makes a promise: if he brings her meat, she will have sex with him that night. And later, after a long day of chasing animals and after a sexually charged ritual feast, she does. In this way, among the Kulina, meat is exchanged for sex.

Other tribes in other parts of the world also value good hunting skills. Studies show that in hunter-gatherer societies, able hunters attract younger and more hardworking wives and tend to have more children than less successful ones. If, as is probable, that was the case in the Paleolithic, too, then such prolific fathers would have been more likely to pass their meat-eating and hunting traditions all the way down to us.

Nevertheless, according to some anthropologists, the role meat played in shaping humanity went far beyond sex and politics. In the words of anthropologist Henry T. Bunn: "Meat made us human."

Most paleoanthropologists would likely agree that meat eating played a crucial role in turning hominins of the past into the humans of today. It's likely the reason why we evolved in Africa and not on another continent. East Africa had an important advantage over other places: because of its climate and rich volcanic soils, it offered our ancestors an abundance of medium-sized carcasses to scavenge. In North and South America, the most available meat was packaged into very large animals, which had few predators, died infrequently, and once dead, left carcasses that were difficult to manage—cutting up a mastodon is not an easy task, after all. In Australia, on the other hand, animals were small, meaning that after a carnivore's feast there would not have been enough meat left to fill hominin stomachs. And without meat, scientists say, we wouldn't have become as big brained as we are today.

Humans have impressively big brains relative to their body size. African elephants, the largest land mammal, are about fifty times heavier than an average American man, but their brains are only about 3.4 times heavier than human ones. The path to our braininess began about 1.5 to 2 million years ago, when the brains of early *Homo* expanded by almost 70 percent in just a few hundred thousand years. Most paleoanthropologists agree that such dramatic increases wouldn't have been possible without a change in diet. Brains are expensive to maintain. Although they account for as little as 2 percent of our body mass, they burn as much as 25 percent of the energy that our bodies need when resting. By comparison, the brains of other primates consume only 8 to 13 percent of the energy needed to run the organism, while those of nonprimate mammals (think mice, polar bears, dogs) use as little as 3 to 5 percent. Our brains are to calories what Hummer trucks are to gasoline—true fuel guzzlers. And yet we don't need to consume that many more calories to sustain these energy-expensive organs. How is that possible?

One widely accepted explanation is simple: something had to give, and that something was the human gut. To be able to sustain large brains without significantly raising our basal metabolic rate, we had to cut costs somewhere else. We couldn't reduce the size of other expensive organs, such as the heart, kidneys, or liver, since that would have made the functioning of our bodies impossible. Instead, it appears that sometime in our evolutionary history, our intestines shrank to make more energy available for the growth of our brains. And that wouldn't have been possible without a better diet.

If you are a *Homo erectus* and you'd like to meet your calorie requirements on a traditional diet of leaves, fruits, grass, and bark, you would need a large gut to digest it all. Such foods are loaded with fiber and need to be eaten in large quantities to satisfy the requirements of a human body. A fruit-eating *Homo erectus*, for example, would need to eat eleven pounds of fruit a day—the equivalent of about thirty-three medium apples. That's a lot of food. Your gut can become smaller only if the things you eat are packed with calories and easy to digest. Peanut butter would have been great, if it were available in the Paleolithic. As it was, our ancestors had to find other high-quality foods to enable the shrinking of their guts and the growth of their brains. That food, most likely, was meat.

It wouldn't take much. It wasn't necessary for early *Homo* to become as carnivorous as saber-toothed cats. If as little as 10 percent of the calories eaten by our ancestors came from meat (about as much as one three-ounce steak on a two-thousand-calorie diet), that would have been sufficient to make a difference between a low-quality and a high-quality diet. Meat was packed not only with calories but also with important nutrients—essential amino acids, iron, calcium, zinc, and sodium, as well as with vitamins A, B1, B6, B12, K, and more.

Was meat the only option? Could our ancestors have improved the quality of their diets without going for animal flesh? Some paleoanthropologists argue that meat alone was not enough to shrink hominin guts: the meat had to be sweetened with honey. Honey is quite the wonder food. It's one of the most energy-dense substances found in nature; it has antimicrobial, antioxidant, antiviral, and anticancer properties. It helps heal wounds and lowers bad cholesterol. If the honey contains bee larvae—as is often the case in nature—it becomes a good source of protein and fat as well. And last but not least, honey is delicious. It comes hardly as a surprise that hominins have been exploiting this resource for millennia. Today, members of some hunter-gatherer tribes such as the Efé pygmies of Africa eat on average 1.3 pounds of honey per day during the "honey season," which runs from July through August. That's 1,900 calories per day in honey.

Another food that has been suggested as "the crucial one" that pushed our ancestors' diets from low to high quality is tubers—those fleshy plant parts deep in the soil, such as potatoes, yams, and Jerusalem artichokes. But though tubers are nutritious and relatively calorie dense, the wild versions can be buried as deep as ten feet down (and so our ancestors would need quite the tools to dig them up), have extremely hard skins, and are difficult to digest.

But perhaps it's not about which new food was added to our diet so much as how we prepared it. Richard Wrangham, Harvard University primatologist, believes that it was not *just* tubers or *just* meat that made us human but *cooked* tubers and *cooked* meat. Wrangham is known in academic circles for his theory that cooking made us human—many scientists I interviewed called him "the cooking guy." Wrangham argues that cooked food is much easier to digest than raw food, which makes us gain weight faster. Since digestion is an energy-costly process—you burn calories to get calories out of food—the faster you can do it, the more energy will be available for your body. In Wrangham's experiments, mice fed cooked meat gained more weight than their fellow cage mates who dined on raw food only.

As for the shrinking guts and growing brains of our ancestors, Wrangham believes that raw meat couldn't have been the instigator. Just look at the dates, he told me. We know that our ancestors were butchering animals 2.5 million years ago, but their brains didn't start to expand significantly until several hundreds of thousands of years later. That's a long gap. On the other hand, if around the time their brains began growing, our ancestors also started to cook their food, that, according to Wrangham, would explain their changing physiques.

There is one problem with the cooking hypothesis, though: fire, or rather the lack of it. Wrangham's critics usually point out that the earliest credible evidence of fire use by our ancestors dates back only to about 790,000 years ago, long after our ancestors' brains expanded significantly. Wrangham replies that absence of evidence is not evidence of absence and sticks to his theory because, he says, there is no other explanation for how humans simultaneously acquired small guts, small teeth, and weak jaws. But no matter the disputes among scientists of whether it was cooked or raw meat that made us human, most agree on one thing: it was meat.

What about honey and tubers? Most likely, by themselves, those weren't the foods that enabled the expansion of our brains. But they did smooth the transition to meat. Hunting and scavenging for meat are energy costly (you burn calories when you fight off lions) and risky (the lions may win). So even though meat was a good resource for making our diets high quality, tubers and honey could have helped when there was not enough animal flesh available. They were a nutritional safety net.

Eating meat did more than simply make the expansion of early *Homo* brains *physiologically* possible. Eating meat, or rather the organized hunting and power scavenging that it required, and sharing the booty were among the important factors that *caused* the increase in the relative size of our brains. The Machiavellian intelligence hypothesis states that we needed bigger brains to deal with the complexities of our social lives: the competition and cooperation, cheating and lying, friendship and play. Meat was a big part of that social life. In a way, it allowed us to even *have* a complex social life. If you eat a low-quality diet of shoots and leaves, the way gorillas do, you need to spend a good part of your day chewing and a good part of it digesting. You're more or less immobile. Neither gorillas nor orangutans are very social animals: there just aren't enough hours left in a day. Eating meat, tubers, and honey enabled our ancestors to reassign time from digesting to socializing.

It wasn't just our brains that grew because of meat; our whole bodies changed as well. Once early *Homo* became meat eaters, they entered the predators' guild. It was a dangerous club to belong to. There were many carnivores back then in Africa, many more than today, and the competition was fierce. From simply being an occasional prey, our ancestors suddenly became rivals, hungry for the same gazelles and antelopes. African predators are known to go out of their way to hunt down competition. In the Serengeti, over 70 percent of cheetah cubs die between the teeth of lions but are left uneaten. Adult cheetahs also sometimes fall victim to lions' competitive streak. It's very likely that Paleolithic lions, saber-toothed cats, wolf-like dogs, and other big carnivores would want to get rid of early *Homo*. Becoming bigger is a good way to avoid being eaten: a time-tested strategy pioneered by the earliest eukaryotic cells. Thus evolution favored larger-bodied hominins, and those that lived, for safety, in bigger groups. And bigger groups led to more Machiavellian intelligence and bigger brains.

The thinning out of our body hair over time also likely had quite a lot to do with carnivory—or hunting, to be precise. Hunting is a vigorous activity. Just think of all the running and spear throwing that need to be done. If it's hot, as was often the case on the African savanna, being covered in thick hair puts you at risk of overheating. That is why, if our ancestors wanted to be good hunters, they needed more adapted bodies. Their hair thinned, and they became better at sweating (for example, we sweat more heavily than our cousins, the chimps—especially on our backs and chests). Once human hair got sparse, the skin got exposed to the fiery African sun. To avoid burns, it became more pigmented, turning darker and darker. And then it came time to move out of Africa: something we might not have been able to do had we not developed a taste for meat.

Finding good food is a big deal when you move to a new country. Whenever I relocate across borders, shopping for groceries becomes difficult, time consuming, even stressful. Something as simple as picking a yogurt off a store shelf becomes a lengthy task of reading labels, comparing prices and ingredients. I have to relearn what I like and relearn which foods taste good, which are healthy and which aren't. Feeding your family well, it appears, takes a lot of knowledge.

When our ancestors first moved out of East Africa 1.8 million years ago, their quest for good food was much more complicated, and eating meat was a way to simplify matters. Many scientists believe that it was our appetite for meat that helped us move out of Africa and spread around the globe. Imagine what would have happened if we didn't eat meat back then and only survived on nuts, fruits, and leaves. Africa is a huge continent, with varied climates and ecosystems. In each of these ecosystems, different plants grow. If you are an outsider, it's hard to guess which ones are edible and which may kill you, making it difficult to move from a savanna you know well to a new environment. But if you are a meat eater, the flesh of one animal is very much like the flesh of another, and in general all mammals and birds are edible. This similarity of potential dinners is one of the reasons why carnivores have larger home ranges than herbivores. Carnivores also need to move around more to find their food. Each day a meat-eating animal will, on average, walk or run four times farther than a plant muncher of a similar size. Meat eating encouraged our ancestors to explore, to step more and more out of their comfort zones. The move out of Africa was not a fast one, though. Our ancestors didn't just pack their suitcases one day and set off north. As Briana Pobiner told me over an ancient cut-marked elephant bone, "When population density increased, groups just started spacing out a little bit more over time to be able to compete for resources." After a few thousand years, some of them ended up in Asia and Europe, and moving into these colder areas really forced our ancestors to establish a meat-eating diet.

Life was far from easy for early *Homo* in Europe. First, there were the plentiful carnivores: wolves, several species of hyenas, cheetahs, pumas, and saber-toothed cats that could weigh up to almost nine hundred pounds. These carnivores were not only happy to eat humans, they also competed with them for food. And in the winter, there was very little else besides meat to fill hominin stomachs. There were no baobab trees, no mongongo nuts, no tropical fruits. Local nuts and seeds, even though highly nutritious, were usually hidden under deep blankets of snow. But there was good news, too. Since European animals carried more fat on their frames to survive the harsh winters, their meat was better for satisfying hunger than that from African animals. If we hadn't eaten that meat, chances are we wouldn't have survived in the frosty landscapes of Europe and Asia.

Yet some of our predecessors, the Neanderthals, took it a bit too far. Neanderthals were good hunters. They chased and killed wild boars, gazelles and deer, brown bears, and wild goats. They were highly carnivorous. Analyses of nitrogen isotope values of Neanderthal bones, which can tell scientists where the protein in the diet of a dead organism came from, show that almost *all* the protein they ate was of animal origin, which would make them almost as carnivorous as wolves or lions. Unfortunately for the Neanderthals, climate change and overhunting drove many species of large herbivores to extinction and decimated others. By the end of the Paleolithic, meat had become harder and harder to come by. Some scientists argue that it was the Neanderthals' dependence on meat that caused their demise. Anatomically modern humans—or moderns, as paleoanthropologists call them—who lived at the same time in Europe and Asia had a more diverse diet. They were less hooked on meat and supplemented the flesh of terrestrial mammals with birds, fish, shellfish, and plants. A diverse diet is better in times of change: if your favorite food is not easily available, you can switch to the second or third best. But if you only know how to eat meat and have no skills to fish or gather nuts, once the animals you usually hunt are gone, you're in trouble. What's more, a diet heavy in meat was quite likely lacking in many nutrients, such as beta-carotene, vitamin E, and vitamin C. This would have meant that the less healthy Neanderthals were likely easily replaced by their better-nourished cousins. And so the omnivorous moderns won, while the Neanderthals went extinct (probably with some direct help from the moderns, too).

For better or worse, meat has played an outsized role in the history of our species. It enabled us to grow bigger brains, encouraged sharing and politics, and helped us move out of Africa and into colder climates. Does that mean we *evolved* to eat meat? That we are hardwired for animal protein and that we should revert to Paleolithic-style diets that are best suited for our Paleolithic-like bodies? Not exactly. "Paleo" may be in vogue, but there are many problems with the premises on which such diets are based. First, the meat you buy today is not the same as the kind our ancestors got on a Paleolithic savanna. The vast majority of our meat comes from domesticated animals who are bred for a high yield of skeletal-muscle meat and are fed artificial diets in confined spaces. This meat has more saturated fat and so is less healthy. A 3.5-ounce strip loin steak from a wild African red deer would have only 0.6 gram of total fat. A similar-sized beefsteak bought in an American supermarket, even an extralean one, would have about twelve times as much fat. What's more, modern meat has a much higher percentage of saturated fatty acids (bad for you) and far fewer monounsaturated and polyunsaturated fatty acids (good for you).

Second, there wasn't one single Paleo diet—either in time or in space. Many of our ancestors were subsisting on leaves, grass, and bark 2.6 million years ago—that's Paleolithic, too. A million years ago, in Africa, early *Homo* likely ate some meat, supplemented with high doses of honey, nuts, baobab seeds, and tubers. Sixty thousand years ago, the Neanderthals ate little but mammalian meat, while some of our other ancestors consumed a lot of fish and seafood. Which of these diets is the "right" one? Which Paleolithic period is more "Paleo"? And why do we even say it has to be Paleo? After all, we've spent much more of our evolutionary time as insect-crunching primates or, later, as fruit-munching apes. Should we all go insectarian or fruitarian, then? Even today hunter-gatherer diets are extremely varied: some are almost vegetarian, while others are heavy on meat. And yet, Paleo diet gurus seem to suggest that there is just one correct way of eating that best suits our bodies. They often give the exact percentages of protein and carbohydrates that you should eat and tell you to stay away from foods such as potatoes, dairy, and cereal grains. They tell you that you *must* eat meat. Why? Because (supposedly) our bodies didn't have enough time to evolve to the new way of eating brought on by agriculture. The problem with this argument is that our bodies *did* have enough time to evolve—and they *did* evolve. The evolution didn't stop ten thousand years ago. According to a growing number of scientists, we have been evolving faster than ever before in the last few thousand years. Studies of the human genome have recently revealed that in the past five to ten thousand years, human evolution has accelerated one hundred times. Several diet-related genes have been identified, genes that evolved after the advent of agriculture. Some of us, it appears, have a new allele (variant of a gene) that regulates blood sugar and protects against diabetes. Others have extra copies of gene AMY1, which helps with the digestion of starches. The most famous example, though, is that of lactose tolerance. Lactose, a sugar found in milk, is indigestible to most adult humans: if you are lactose intolerant and drink milk, you may end up with stomach pains, diarrhea, and even vomiting. Yet lactose intolerance is not common among all nations. For example, in some societies in northern Europe, over 95 percent of the population can digest milk without problems, while only about 1 percent of the Chinese are able to do so. The reason? Early domestication of cattle by northern Europeans and a few thousand years of evolution.

There is one gene, though, that seems to have evolved because of our growing appetite for meat. A rather important gene, too. It's called apoE, and it comes in three basic variants: E2, E3, and E4. If you have the E4 allele (in many labs you can order a genetic test to determine this), your life may be shortened by several years compared to that of someone who has the E2 or E3 allele. If two people, an E4 carrier and an E3 carrier, both add two egg yolks to their daily diets, the E4's blood cholesterol may spike up four times more than that of the other guy's. No wonder that E4 allele carriers have about a 40 percent higher risk of heart disease. So why would nature have us evolve such a lousy variant of a gene? The answer is: to enable us to eat meat. E4 is not all bad news. It developed well before we learned how to control fire and cook our meat. Eating raw flesh is dangerous, especially if it's scavenged and rotten, crawling with parasites, bacteria, and viruses. Wild chimps, for example, get the deadly Ebola virus from eating colobus monkeys. Yet the E4 gene variant, which boosts the immune response of the body, enabled our ancestors to eat such tainted meals without getting sick too often. Unfortunately, it also put them at risk for aging faster. That's why a few hundred thousand years later, another meat-adaptive mutation happened, and E3 appeared on the scene. From then on, carriers of this new allele could eat fattier meat without too much risk to their hearts.

Today, some of us are still left with the meat eaters' version of the gene. Among Americans of European origin, about 13 percent are carriers of this life-shortening allele. The newer E3 is the most common allele worldwide, especially in Japan, China, and India. Does that mean that carriers of these meat-adaptive genes have a particular taste for animal flesh, or that they need meat more than others do? Not at all. Actually, the carriers of the E4 allele would be better off on a low-fat vegetarian diet (although if they wanted to try some rotten meat, they would be more likely to survive this adventure than the carriers of E3 or E2).

We shouldn't assume that just because a diet is "ancient" it must be good. As one writer joked in an article for the *Canberra Times*: "Try the latest Paleo diet and you too can be short, stocky, hairy and smelly and . . . then you die." Cavemen didn't exactly lead idyllic lives. Their preserved skeletons tell us they suffered from arthritis, gum diseases, deformed limbs, and cancer. Sure, our modern diets are often far from balanced (too much junk food, too much sugar), but they can also be unusually good (veggies from across the globe, fruits year-round, all the seeds and nuts you can imagine). We have options to eat better than our ancestors ever could and ever did. What history can definitely teach us is that we are highly omnivorous and highly adaptable and that we can thrive on many different diets—even rather extreme ones.

Would we have become human had we never developed the taste for meat? Would we have become the big-brained, hairless, social creatures we are today, living across the globe from Europe and Asia to the tiny Pacific islands of Tonga and Nauru? Maybe. Our ancestors didn't need meat per se to evolve from the hominins of 2.6 million years ago to *Homo sapiens* of today. Meat was not a physiological necessity. What they did need was a high-quality diet, and at the time meat was the best option they had. That's why they got hooked on it. Maybe they could have chosen something else. Maybe they could have gone for baobab fruits, which are loaded with proteins and other nutrients. Maybe they could have invested more time in "hunting" insects. After all, chimps can meet their daily protein needs in as little as thirty minutes of termite fishing. Maybe they could have just eaten more honey, tubers, and seeds. All this would have made for a high-quality diet. But, in a way, meat *was* special. Only meat was both highly nutritious and dangerous to come by (which led to male showing off and politicking). Only meat came in big enough packages to encourage sharing. Only meat required chasing after, which led to us losing our heavy coats of hair. Only meat was basically the same across continents.

What the history of our meat eating can teach us is that our ancestors were highly adaptable. We are not meat eaters by nature so much as we are opportunists. Our ancestors changed their diets quite dramatically several times in the past—from insects to fruits, from fruits to grasses and leaves, from grasses and leaves to meat and tubers—usually in response to changes in climate. Fruits were best for them at some points in the past; meat was best at others. Rather than dumping peanut butter and potatoes down the drain, we should take another lesson from our Paleo (and earlier) ancestors: instead of looking for a perfect and "natural" diet from the past, start looking for one that would be best for right here and right now.

It's also important to keep in mind that although our ancestors did revolutionize their diets a few times, it doesn't mean it came to them easily or quickly. After all, they didn't alter their eating habits over a year or two—it took thousands, at least. Make the change fast, and we may resist. And if the change means less meat, we may resist even more strongly—acting as if without animal flesh on our plates we are doomed to wither and die.

So a new question pops up: Is there some compound in meat that actually helps our bodies function better? Something that, if removed, would mean worse health and less sparkling brains? A vitamin or a macronutrient capable of enticing a powerful hunger-like craving? To answer that question, we must turn from paleoanthropologists and their boxes of bones to biochemists with microscopes and find out meat's molecular secrets.

3

The Good, the Bad, and the Heme Iron

The line was about thirty people long, snaking out of the store, spilling onto the sidewalk. At least two hours would go by before the person at the end of the queue could make a purchase. But they all stood patiently, shuffling a few steps ahead every four minutes or so. In their hands, they clutched polyester bags and wicker baskets, which they hoped to fill with meat. And yet, it was likely some of the shoppers would have to go home with nothing. The butcher's store was a white, tiled expanse of emptiness. Only a few sad sausages dangled on a line of hooks sticking out of the wall. Everybody wanted those sausages.

During my childhood in Poland in the early 1980s, such scenes were common. Just like everyone else, my mother and I would wait for hours for a chance to buy some pork or beef. Meat was rationed back then; it was scarce and voraciously craved. I doubt my mother, or anyone else I knew, would sacrifice so much time to buy a few ounces of beans or a head of cabbage. But for meat, we waited.

The late anthropologist Marvin Harris was so fascinated by this Polish desire for sausages and schnitzels that he used it as a prime example of what he called "meat hunger"—a universal human craving for animal flesh that can't be satisfied by any other foods, no matter how plentiful. In the 1980s most Poles were far from malnourished. We ate over three thousand calories per person and loaded up on over one hundred grams of protein a day. Still, we would spend drudgingly long hours queuing in butcher shops. Why were we so hooked on meat?

Harris may have been the most famous believer in meat hunger, but he wasn't the first. Back in the nineteenth century, missionaries and explorers wrote quite extensively about a condition they encountered in Africa and South America: no matter how abundant their food, the locals would complain of being hungry if they didn't have meat to eat. In 1867 one French American traveler described a "disease" prevalent in Central Africa called *gouamba:* "An inordinate longing and craving of exhausted nature for meat." A person touched by *gouamba* would refuse any vegetarian dishes brought before him and stubbornly beg for meat.

In many languages there actually exists a word for "meat hunger" to show that it's a different thing from the regular, empty stomach type of hunger. It's called *ekbelu* by the Mbuti tribe of Central Africa and *eyebasi* by the Yuquí of Bolivia. The Mekeo of New Guinea say that "hunger for plant food" comes from the abdomen while "hunger for meat" comes from the throat. In Uganda, locals have been known to exchange plantain that could feed a family for more than half a week for a bony chicken that won't last them even a day. Anthropologists say that meat hunger is not physiological but rather cultural—a lack of animal flesh in the diet is a sign of scarcity, a symptom of problems with a tribe (or nation). And yet, there may indeed be something about meat's nutritional value that makes us prize it above such foods as plantain, something that makes meat hunger real. That thing is protein.

Paul Breslin's fly lab at Monell Chemical Senses Center in Philadelphia smells of honey. It's a small room, with little more inside than a desk, a computer, and a big, white fridge. In here, Breslin, a professor of nutritional sciences at Rutgers University in New Jersey, studies the feeding behavior of an animal species that is particularly fond of wine, beer, and freshly baked bread. And no, I'm not talking about humans—Breslin studies fruit flies. According to Breslin, fruit flies' appetites are very similar to that of humans and even more similar to that of chimps—at least tastewise. "They practically *are* chimps," he jokes, as he pulls open the fridge to reveal rows of transparent vials, each buzzing with tiny, orange-hued flies. The vials have a hardened, gooey substance at the bottom—fruit-fly food, Breslin tells me (hence the honey-ish smell). By studying fruit flies and mosquitoes, Breslin seeks to understand what drives their protein hunger and, in turn, what may be causing our own desire for protein. "These flies eat little but fruits, but once they get pregnant, the female's protein hunger turns on, and she starts looking for yeast—that's a source of protein for them," Breslin says. It's similar with the mosquitoes. Breslin believes that if it wasn't for the itchy bites (annoying) and diseases such as malaria (deadly), we would smile at any mosquito trying to get our blood. After all, the female mosquito is steered by her protein hunger: she is pregnant and trying to get enough of the nutrient to feed her babies.

Mosquitoes and fruit flies aren't the only creatures driven by their desire for protein. Stephen Simpson, professor of biological sciences at the University of Sydney, has studied slime molds (jelly-like creatures that resemble toxic spills), cockroaches, rats, monkeys, and humans and discovered that all share a special craving for protein. Simpson told me once that for people the desired ratio of calories from protein in the diet is about 15 percent. Whenever in his experiments Simpson offered volunteers diets too low in protein, they ended up snacking on savory foods, as their bodies told them to meet their "protein target." Since it's usually the taste of salt and umami that signals the presence of protein, Simpson listens carefully to what his own body is trying to tell him. If he finds himself craving salty potato chips between meals, he thinks, "Oh, here comes the protein hunger again"—and eats an egg instead.

It appears that our bodies are designed to prioritize protein and actively seek it out—whether you're a mosquito sucking blood and spreading malaria or a human scavenging the fridge for bacon. From this perspective, the Ugandan family exchanging plantain for a bony chicken makes sense. Plantain is quite energy dense but has very little protein: about one gram per one hundred grams of fruit. To meet daily protein needs, an adult man would have to gorge on thirty pounds of plantain in a single day. Meanwhile, the scrawny chicken has about twenty times more protein per one hundred grams than the plantain and would satisfy the protein hunger much more efficiently.

Yet Simpson's and Breslin's research can't explain such cravings completely. Though they have helped explain why humans may sometimes fall victim to meat hunger, they haven't shown that we have an innate hunger for meat per se—just that the human body craves protein. It also doesn't mean we need to cram ourselves with as much protein as our belts will allow—15 percent of calories from protein is actually not that much. The beliefs that so many Westerners hold—that meat equals protein and that our bodies require vast amounts of the nutrient—are nothing more than a myth. Or two myths, to be precise, both of which have roots in science.

In 1824, Justus von Liebig became a professor at the University of Giessen, Germany, at barely twenty-one years old. Von Liebig's career was not only fast but also quite impressive: for his discoveries of the role of nitrogen in plant nutrition, he is still known as the father of the fertilizer industry. Some of von Liebig's fame was due to his research, and some of it due to his charm: he was a great self-promoter. And that definitely helped spread his ideas about nutrition and meat.

Von Liebig glorified protein as the only real nutrient and believed that without it our muscles just wouldn't work. To him, carbohydrates and fat were only needed to react with oxygen in the lungs to produce heat. The fact that his ideas about the role of proteins in human nutrition were purely speculative and not based on experiments didn't deter him from turning them into a business opportunity. His "Liebig's Extract of Meat," an essence made of Uruguayan beef, was widely marketed as a "powerfully-acting panacea" capable of restoring health. According to von Liebig, meat was what people needed to be strong, and since he was a famed scientist, many believed him.

It wasn't only von Liebig, though, who was behind the protein myth that took hold in nineteenth-century Germany and spilled from there all over the world. It was also his student, Carl von Voit. Although von Voit did conduct quite a few experiments that have helped us understand proteins better, his nutrition advice was based on rather shaky science. He calculated how much protein soldiers, laborers, or prisoners consume each day and from this inferred that the resulting number represented how much their bodies actually need. The problem with his methodology is obvious: it's a bit like observing children stuffing themselves with cookies and concluding that young humans require tons of sugar to grow.

Von Voit advised that hard workers should eat a staggering 150 grams of protein a day and that at least 35 percent of the nutrient should come from meat. The idea that we need plenty of protein swiftly became popular among nineteenth-century elites, most of them devout meat lovers already, and soon even scurvy was blamed on inadequate protein intake.

The protein myth had decisively taken root. But in coming years, well-designed scientific experiments cast a long shadow of doubt over the German scientists' claims. By 1944 the US Department of Agriculture (USDA) recommended only seventy grams of protein per day for adult men and sixty grams for women, "regardless of the degree of activity," and von Liebig's and von Voit's prescription for excessive protein began fading away fast. It may have disappeared altogether if it wasn't for kwashiorkor.

In Ghana *kwashiorkor* means "evil spirit that infects the first child when the second child is born." It's not an evil spirit, of course, but a disease that commonly falls upon an older sibling taken off a mother's breast so that a newborn can be fed. Once the first child is weaned onto a starchy, nutritiously inadequate diet, he may develop swelling of the legs and the ill-famed bloated belly and become susceptible to infection. Observations in Uganda in the 1950s led many scientists to believe that the disease was caused by too low an intake of protein. Soon the world fell into what some now call a "protein hysteria," with charities and governments worrying about the "protein gap" between the rich and the poor countries and sending skim milk and fish flour overseas to boost the protein intake of hungry African children. But the truth was, as scientists finally admitted in the 1970s, that kwashiorkor was more about inadequate dietary intake than about protein. If your total dietary intake is too low, your body will burn whatever protein you eat for energy, instead of using it to make your own proteins—like insulin, collagen, or antibodies.

In general, diets that are sufficient in calories will also provide enough proteins. Of course, it is possible to imagine a diet loaded with calories but with barely any protein in it: a cotton candy regime, for example. But if your menus are fairly balanced, you should be fine, even if you are eating vegan. What's more, the reason why we don't see many people with kwashiorkor walking the streets of upper Manhattan is that it's extremely hard to succumb to this condition in the West unless you are truly starving, an AIDS patient, or a drug addict. Detailed studies of diets in developing countries showed that even there the lack of protein itself was rarely a problem; there just wasn't enough food in general. Once children were given a diet with sufficient calories, the intake of protein in most cases was adequate. It was soon discovered, too, that the protein requirements of children were previously deemed too high. They were calculated based on experiments on rats, but nonprimate animals grow much faster than human babies do and so require more protein. This is reflected in the protein content of rat breast milk, which is much more loaded with the nutrient than human milk is. (Compared to other mammals, human breast milk is *very* low in protein.) Once that was corrected for, children's protein requirements were slashed to less than half of what they were in the 1940s. The protein gap practically disappeared, but the protein myth persisted.

If you leaf through today's consumer magazines, especially those intended for men, or if you browse the nutrition pages of the Internet, you will soon discover that the protein myth is still going strong. "Simply put, our muscles are meat, so we need to eat muscle to gain it," claims *Flex*, a bodybuilding magazine. "Muscle is made of protein, so to grow lean mass you need to eat protein. Ample protein," warns an article in *Muscle & Fitness*.

So how much protein do we actually need? According to the Centers for Disease Control and Prevention (CDC), the recommended dietary allowance (RDA) for protein is 0.8 gram per kilogram of body weight per day. It's the same for all adults, both women and men, and covers everyone from couch potatoes to gym bunnies. The RDA (just like the British Reference Nutrient Intake) is so designed that consuming this amount of the nutrient is more than enough for 97 percent of the population. So it will cover the needs not only of a regular adult but also those of people with significantly elevated protein requirements, such as those with cancer, for example, or other serious diseases. The protein requirement of an "average" American, according to the CDC, is just 0.66 gram per kilogram of body weight per day.

This is not to say that proteins are not important. They are, and you need to eat them. What you don't need to consume, though, is animal flesh. The second myth is that protein equals meat, and that by going meatless, vegetarians are somehow endangering their bodies and minds. This particular myth goes deep into human history, but in one of its modern incarnations, it can be traced back to a best-selling 1971 book, *Diet for a Small Planet* by Frances Moore Lappé. The myth starts with a correct assumption that not all proteins are created equal: some are just better for you than others. It's all down to their amino acid composition. Proteins are built from about twenty common amino acids that can be assembled together like chains of beads. When you eat a protein-rich food, your body breaks the protein down into separate amino acids (think of beads from a broken necklace, scattered on the floor) and then rearranges them into your own, different proteins (a new necklace). Your body can synthesize a few of the amino acids by itself and doesn't need to get them served on a spoon: these are called nonessential. Other amino acids are essential, and you have to get them with food. Some proteins have all the nine essential amino acids, and for this inherent goodness are called complete proteins. Egg proteins are like that, or the ones found in meat. But most plants lack several of the essential amino acids; so for example, if you eat nothing but beans, you won't get methionine, and your body functions will soon start to fail. This may sound scary if you are a vegetarian, but the good news is that all essential amino acids are found in plants, spread among different veggies, fruits, and grains. So even though beans lack methionine, you can get this amino acid from grains, making a black bean burrito a perfect protein match. Another classic combination is a peanut butter sandwich. But what if you don't eat grains and beans at the same time? Here is where Frances Moore Lappé comes in. When her book was originally published in 1971, she wrote that people should combine plant amino acids on their plates to create foods equal in their protein goodness to meat. If vegetarians were careful and charted the essential amino acids to assemble their foods, they would be just fine.

The problem was the charting, of course. You had to be aware of which plant foods are short in which essential amino acids and plan your meals accordingly. It wouldn't do to polish off a spoonful of peanut butter at 9:00 a.m. and have a slice of bread at noon. You had to eat them *together.* Vegetarian books started printing charts of complementary proteins that were supposed to help, but mainly just added ammunition to meat eaters' claims that plant-based diets are a disaster waiting to happen.

We now know that vegetarians don't need to chart their amino acids any more than omnivores need to chart their vitamins to make sure they get all of them at every meal. A human body is perfectly capable of complementing the proteins by itself. Lappé admitted in the twentieth-anniversary edition of her book: "In combatting the myth that meat is the only way to get high-quality protein, I reinforced another myth . . . Actually, it is much easier than I thought."

Not only is it unnecessary to chart your amino acids, there are some plant foods that have high-quality protein just like meat. Soy is one example, as are buckwheat, quinoa, and even potatoes. If you went on a potato-only diet (fries for breakfast, purée with chips for lunch, and potato pancakes for dinner), your body would get all the essential amino acids it needs. About three pounds of potatoes a day should do the trick.

If you are a Westerner, you most likely don't have to worry about eating enough protein—just the opposite. People in developed countries tend to pack in far too much of the nutrient. Americans eat about twice as much as they should. Even professional Olympic athletes generally don't have to supplement their meals with additional protein. Their protein requirements are at best only a little higher than those of the general population: 1.2 to 1.7 grams per kilogram of body weight per day. Sportspeople burn tons of calories. Scott Jurek, one of the world's best ultramarathoners (and a vegan), once told me he has to down about six thousand to eight thousand calories each day to make up for all the exercise he does. Because of such elevated food intake, professional athletes can easily meet their protein needs with their regular diets. If you go to the gym a few times a week, by the way, you basically still only have to get 0.8 gram per kilogram of body weight per day, like everyone else. As Breslin told me in his fly lab: "Humans who want to look like Schwarzenegger eat a lot of protein to grow their muscles, but they'll never look like gorillas, never be as muscular, even though gorillas don't eat meat."

What's more, too much protein is not just bad for you, it may possibly kill you. By 1919 Vilhjalmur Stefansson, a Canadian explorer of Icelandic origin, had spent a total of ten winters and twelve summers north of the Arctic Circle. Sometimes for as long as a year and a half, he would live in the frozen expanse of northern Canada on nothing but wild game that he managed to hunt down with his rifle. He survived with no fresh veggies, no bread, no tea, or even a pinch of salt. Although Stefansson believed that humans could live well on meat without vegetables (he actually managed to gain fifteen pounds on one of his trips), he also described the dangers of very lean, very protein-rich meat: "If you are transferred suddenly from a diet normal in fat to one consisting wholly of rabbit you eat bigger and bigger meals for the first few days until . . . you are showing both signs of starvation and of protein poisoning . . . Diarrhoea will start in from a week to 10 days and will not be relieved unless you secure fat. Death will result after several weeks." This condition, called rabbit starvation, was reported by many travelers in the nineteenth and twentieth centuries. Even though there weren't any actual experiments done on rabbit starvation in humans (it would be rather hard to find volunteers, I presume), quite a few studies show that it can be dangerous if over 35 percent of the calories a person eats come from protein. If you are a 110-pound woman, for example, just 125 grams of protein a day may be too much for you, an amount that may be exceeded in a single McDonald's visit (the Mighty Wings, ten pieces, have 60 grams of protein). What happens if you overindulge on animal protein? Your kidneys may stop functioning properly, even fail if you are a diabetic. A high-protein diet may also set you on the way to developing osteoporosis, heart problems, and cancer. Studies show that mice fed a diet with a high protein-to-carbohydrate ratio have a shorter life span, especially if the protein comes from animals. In a more prosaic scenario, you can suffer from constipation. Up until the mid-nineteenth century, it was considered a national affliction among Americans, precisely because of a meat-heavy but veggie-light diet.

So if meat is not necessary in the diet for its protein content, is there something else in it that does make it indispensable or at least much better than plant foods? Some important nutrient that keeps our bodies craving for more? When you read about the adequacy of vegetarian diets, usually a list of vitamins and minerals is provided as the ones to watch out for: iron, vitamin B12, zinc. Could it be that one of these nutrients, if lacking, turns the meat hunger on?

If anything, iron would be the most likely candidate, since not all iron in food is created equal. Some of it is in "heme" form (derived from hemoglobin, the protein that makes blood look red) and can only be found in animal products: meat, eggs, fish. The rest is nonheme iron, again found in meat, eggs, and so on, but also in plants—particularly in beans, spinach, and nuts. If you've heard about heme and nonheme iron, you've probably also heard that heme iron is much more easily absorbed by the body. You may have heard that for this reason heme iron is better for you. But more and more research shows that heme iron may actually be the bad guy, and that anemia can sometimes have benefits (yes, that's right).

What's wrong with heme iron? It may promote cancer and cardiovascular disease. Many of the studies headlined in the media, which connect red meat consumption with these conditions, point to heme iron as the likely culprit. As for anemia, it's not always due to inadequate consumption of iron. Even though anemia is the top nutritional deficiency in the world—almost half of children and women in developing countries are anemic—sending them steak and bacon would not automatically solve the problem. Studies show that not enough iron in the diet is less of an issue than are genetic disorders, chronic inflammation, and worms. Among Tanzania's schoolchildren, for example, 73 percent of incidents of severe anemia were brought on by hookworm infections (that's because the creepy-crawlies cause intestinal bleeding). Besides, anemia itself is not necessarily all bad, either. Evidence is starting to pile up that anemia can sometimes protect us from succumbing to infectious diseases, such as malaria, tuberculosis, or even HIV. Most bacteria and viruses need iron to survive in a human body. If there is a shortage of the nutrient, they can't multiply as successfully and overwhelm the immune system. It has been suggested that anemia might be an adaptation that allows us to live in places where infections are common. Meat-scarce diets could have actually helped our ancestors survive the unhygienic world of the past.

Today, Western vegetarians are no more likely to have anemia than are omnivores. Even though nonheme iron is less well absorbed by the body, contemporary plant-based diets are more abundant in this nutrient, making up for the difference. And it's not just spinach and broccoli that can load a vegetarian with iron. Keeping anemia at bay can actually be quite pleasant; one good source of iron is the candy licorice allsort (at eight milligrams of iron per one-hundred-gram serving, it has nearly four times as much iron as a three-ounce sirloin steak). Of course, in the past the story was significantly different. If you lived in a medieval European village circa 1300 and ate little but turnips, your risk of anemia was much higher, and a steak would have helped.

Some studies do find, though, that even modern plant eaters may have lower iron *stores* than omnivores do. That's like having less food stacked in a pantry: you're still fine (no anemia) but more likely to run short in the near future. But low iron stores are not necessarily a bad thing. First, if someone has low iron stores, his or her body will be more efficient at absorbing the nutrient from meals. Second, low iron reserves may be why premenopausal women have lower risk of cardiovascular disease than do Western men.

What about zinc, then? Is that the nutrient that keeps us hooked on meat? After all, zinc from plant foods is not as well absorbed as the one from meat. But, once again, the answer is no. Study after study shows that vegetarians do not have any zinc deficiencies, no matter what meat producers would like us to believe.

However, there is one nutrient that modern, Western humans tend to only get from animal foods: vitamin B12. There simply are no plant sources of this compound. Kelp, tempeh, and miso, which are all often proposed as great sources for getting B12, contain only inactive analogs of the vitamin and won't keep deficiencies at bay. The only places to get B12 are meat, eggs, and dairy products. And if you don't get it, your nerves won't work as they should, and you won't be able to make healthy blood cells.

If you wonder how our hominin ancestors survived the Paleolithic without any milk, eggs, or meat in their diet (as was, after all, common) and didn't die off like flies from B12 deficiencies, the answer is simple:

dirt. Vitamin B12 in meat doesn't come from the animal itself; it comes from microbes. It's produced exclusively by soil bacteria living among the roots of plants and ends up in meat when an animal eats dirty grass, leaves, fruits, and so on. That is why in developing countries B12 deficiencies are not as prevalent: people there don't wash and spray their veggies as vigorously as Westerners do. Yet it doesn't mean that every American vegan should dash to the nearest ditch and gobble up soil three times a day. A supplement will do just fine, or consuming enough B12-fortified foods, such as cereals and soy milk. For vegetarians their regular intake of eggs and dairy is normally enough to keep their bodies well supplied with B12.

It seems, then, that it's not vitamin B12 that keeps our bodies dependent on meat, just as it's not protein, heme iron, or zinc. But what if there is some compound in meat that we haven't discovered yet, some nutrient that, say, in 2030 scientists will finally pin down, and all vegetarians will have to bow their heads in embarrassment and admit: the omnivores were right all along; our bodies *do* need meat.

Even now some more obscure compounds are being proposed as the ones behind meat's nutritional superpowers. One study presented at a 2013 congress in Turkey (which was, by the way, sponsored by the meat industry) suggested carnosine and anserine, antioxidative peptides that are only found in meat, as substances that may boost our health. But the fact is, that no matter how many "new" compounds are put in the limelight, it's highly unlikely that we will ever find anything in meat that makes it necessary for our bodies, or even much healthier than nutrients present in plant foods. Why? Because of all the longitudinal studies that have already been conducted comparing the overall health of vegetarians and meat eaters.

During World War I, an unusual natural "experiment" happened in Denmark. By January 1917, a naval Allied blockade of the country (which was then occupied by Germany) resulted in severe shortages of grain and fertilizer. As the months passed, the threat of hunger loomed larger and larger. Mikkel Hindhede, one of Denmark's top nutritional scientists, whom some called the "joy-killing missionary," proposed a drastic move: slaughter most of the country's pigs and divert the scarce grains directly for human consumption. His advice was promptly followed. Overnight, practically all of Denmark went vegetarian. They subsisted on little but rye bread, barley porridge, potatoes, green veggies, milk, and some butter. Not only did Danes escape hunger, but within a year, the death rate from disease fell by 34 percent. Later, that period from October 1917 to October 1918 was dubbed the "Year of Health." Of course, the benefits may have been due to other factors, too: the Danes drank much less beer as well, for example. But if meat was truly so indispensable, most likely more people would have died during the blockade, not less.

Many studies show that vegetarians have lower mortality rates than omnivores and are less likely to succumb to cancer or heart disease. The vegetarian Seventh-day Adventists in California, for example, live on average 9.5 (men) and 6.1 (women) years longer than other Californians. "Nutrition experts have known for decades that plant-based diets provide more than enough protein. In our studies, we consistently find that, as people switch from an animal-based diet to a plant-based diet, their diet becomes richer in vitamins, fiber, and other important nutrients. There is never a need to add animal products," says Neal Barnard, professor of medicine at George Washington University, who has conducted numerous studies on plant-based nutrition.

Of course, anyone can have a horrible diet, and that applies to both plant eaters and meat eaters. A person surviving on nothing but fried bacon and pepperoni pizza would soon end up with vitamin C and K deficiencies, for example. A fruitarian diet can be dangerous, just like a macrobiotic diet, or those proposed by Robert Atkins and Pierre Dukan (the British Dietetic Association says they are not "nutritionally balanced").

But just because there is nothing in animal flesh to keep us nutritionally hooked and we, in the twenty-first-century West, don't need it to be healthy doesn't mean, however, that this has always been the case. It's not even true today: in some places on the planet, such as the Arctic, the only food available may be meat. Consider such a metaphor: If you are hiking in the desert with little water available, then drinking out of a dirty creek swarming with bacteria is a good idea, and you should do it in spite of the health risks (dying of dehydration is not very healthy). But if you are hiking with a whole backpack of Evian, then filling your cup in the creek is not only unnecessary but a rather bad idea. Eating meat is like drinking that creek water: it may be absolutely vital at times but not necessarily the best choice if you have great plant foods around.

The premodern diet for the vast majority of the world's population was poor. There were very few vegetables to choose from, few grains to cook, and, in the North, no vegetable fat to provide calories. There was often not enough protein to satisfy hunger. People needed meat because they couldn't have a rich diet otherwise. That's where our stereotype of a sickish, bone-thin vegetarian may come from: most of those who were vegetarian in centuries past didn't eat meat (or anything else) because they were too poor and so were often hungry and weak. This heritage of "you need meat to survive" may have gotten engraved in our culture as a powerful myth that lives on and keeps us craving animal flesh.

There is another theory, though, that may help explain why human beings desire meat rather than beans, tofu, or spirulina. According to this theory, it's all due to our so-called selfish genes. Evolution doesn't necessarily favor those who live the longest; it favors those who can reproduce the most. Studies show that the more animal protein people consume, the sooner they become fertile and the more kids they can have. Girls on meat-rich diets get their first periods earlier than their vegetarian counterparts. If the difference is as big as three or four years, as some studies suggest, it means meat eaters can have a few more kids each. The fact that these prolific parents may die earlier of cancer or heart disease doesn't matter to the selfish genes.

Besides making us fertile earlier, it's highly unlikely that meat keeps us nutritionally hooked. There is nothing vital in beef or pork that plants can't provide. We don't love eating meat because we have to do so to stay healthy. Yes, meat can satisfy our protein hunger quite well, but so can a peanut butter sandwich. Yes, meat is a great source of iron but so is licorice. There is nothing in meat that the Polish nation can only get from sausages and schnitzels, nothing to justify (from the nutritional perspective) the long waits we endured in the butcher stores of the '80s.

Yet for the many tribes of Africa or South America, meat hunger may be very real. If there is no other protein around, meat can help people reach their protein targets. The fact that in the past we so often *did* need meat for its nutrients because there was little else to eat, and that animal protein is good for our reproduction-happy selfish genes, may have left us with taste buds sensitive to the particular mixture of compounds found in animal flesh. Westerners may not need meat anymore to live healthily, but our tongues and noses obviously didn't get the memo. They still make us crave meat's perfect brew of flavors—of umami, of fat, of the products of the Maillard reaction—keeping us hooked, even if eating meat may be against our best interests.

4

The Chemistry of Love: Umami, Aromas, and Fat

The flat, car-dotted landscape between Wharton Street and East Passyunk Avenue in south Philadelphia is scorched by the sun. It's hot, it's humid, it's loud with traffic. It is also the center of Philly's cheesesteak kingdom, a small patch of the city where all the best cheesesteak restaurants are located. There is so much steak grilled around here that the air itself smells of beef.

I'm here because I'm going to eat meat—for the sake of science. I feel guilty about the fact that I'm about to swallow a small part of what once was a living, breathing cow, but I am also quite excited: the cheesesteaks are internationally famous as among the best meat-based dishes America has to offer, and I am here to find out what makes cheesesteaks, and meat in general, so appetizing.

Pat's King of Steaks is a simple sandwich place with a clutter of red tables and benches that spill onto a sidewalk. It was founded in the 1930s by Pat Olivieri, who claimed to have invented the first ever Philly cheesesteak—a mouthwatering combination of melted cheese on thinly shaved grilled meat in a soft bread roll. President Obama ate here, and so did Senator John F. Kerry, as well as a plentitude of celebrities, whose pictures now line the walls.

Today, there is a long line of customers, each ordering swiftly and with confidence, *wit* onions and *wit* cheese (*wit* means "with" in

Philadelphese). When I take a big bite of my sandwich, the full-bodied flavor of beef instantly fills my mouth. The meat is fatty, packed with aromas. It's good, very good. I can truly see (or rather taste) what all the fuss is about. There's something about the flavor of cheesesteaks that can be truly enticing, even to someone who doesn't eat meat.

When prisoners on death row in the US ask for their last meals, their most common request is, by far, for meat. Studies show that 74 percent of American men and 61 percent of women crave meat the way others crave chocolate or ice cream. We've long known why ice cream and chocolate appeal so much to our palates: it's that blissful mixture of sugar and fat. But what's so special about meat that makes our taste buds beg for more? What is it about bacon and steak that attracts us, despite the growing pile of scientific data on meat's detrimental health effects (think cancer, heart disease, type 2 diabetes)?

When I mention my cheesesteak experience to Gary Beauchamp, professor of biopsychology and an expert on taste perception, he actually chuckles. "Oh, yes, cheesesteaks. These are *so* good," he says. Beauchamp is the former director of the Monell Chemical Senses Center in Philadelphia, where a good chunk of the world's research on taste and smell takes place. It's the same institution that houses Paul Breslin's fruit-fly lab. That's not a mere coincidence. Although the low-ceilinged labs of Monell have a rather '70s-like feel, the research that is being conducted in here is very twenty-first century. The names of the scientists that adorn office doors and are scribbled on chemical vials and on the cages of lab mice (to make sure no one messes with other people's work) are the same ones that constantly pop up on the front pages of science sections in newspapers and magazines.

Beauchamp's office on the first floor is cheerfully cluttered with all the usual cliché scientific paraphernalia: papers stacked high on the desk, cardboard boxes full of jars, ethnic masks hanging on the walls. Beauchamp himself looks like the stereotype of a distinguished scientist: he's slim, silver bearded, and soft smiled. And he loves talking about research.

When I ask him what makes meat so attractive to our taste buds, Beauchamp thinks for a while, then replies: "It's not just one thing but a combination of things. Meat is rich in umami; it has a lot of fat. And this umami and fat combination seems to be a highly interactive one."

Beauchamp believes that the key is meat's unique mixture of umami and fat that gets spiced up when meat gets browned during cooking—a process called the Maillard reaction that produces particularly desirable flavors and aromas. How you perceive this mixture, though, depends on your genes, your sense of smell, and the number of tiny, mushroom-like structures on your tongue called fungiform papillae, on which taste buds are perched.

We can detect five basic tastes: salty, sour, bitter, sweet, and umami. Some scientists argue, with growing success, that we can also detect the taste of fat, while other researchers (with less success) are trying to prove the existence of calcium or metallic tastes. Some even say we may be able to perceive such tastes as electric and soapy, but data on that are *very* thin. What we do know is that the tongue or taste map that originated in 1901, which you may have stumbled across in old science textbooks, is incorrect. According to that map, we are able to detect specific tastes only in specific regions of our tongues. That is not true; in fact, all tastes can be detected almost anywhere on your tongue. It happens like this: you slide a piece of food (a slice of bacon, say) into your mouth. The saliva moisturizes the food and helps release the molecules of substances that evoke tastes, such as sodium chloride, which is responsible for the salty taste, or monosodium glutamate (MSG) for umami. These molecules then float toward your taste buds and bind to specific receptors on their surfaces. Later, three main nerves in the head called cranial nerves carry the taste messages to your brain, which make you either cringe with disgust or crave for more. Bitter and sour tastes are generally a warning that the food in your mouth may be poisonous or spoiled. Sweet means full of carbohydrates and calories (good), salty means sodium (necessary for proper functioning of our bodies), and umami most likely means proteins.

When it comes to perceiving flavors—responding to the tastes, textures, and aromas of foods—we are not all created equal. We have different densities of fungiform papillae and, as a result, different numbers of taste buds. Some of us have only two thousand in total; some have as many as eight thousand.

The discovery of why the number of our fungiform papillae matters to our taste perception is a rather recent one and dates back to a puzzling laboratory episode in 1931. It was then that Arthur Fox, a scientist working for DuPont chemical company in Wilmington, Delaware, accidentally spilled a substance called phenylthiocarbamide (PTC) in his lab. When his colleague, C. R. Noller, complained about the awful-tasting stuff floating in the air, Fox was puzzled—he couldn't smell anything, much less taste it. To prove Noller wrong, he put some of the white powder on his tongue—and found that to him it had no taste whatsoever. This exchange prompted Fox to study the taste of PTC. Soon he discovered that while some people, like Noller, find the substance unbearably bitter, others, like himself, can't detect any flavor at all in the powder. Fox calculated that about 28 percent of people can't taste the bitterness of PTC. What Fox didn't manage to do is figure out *why* some of us are supersensitive to PTC and some of us aren't. It was Linda Bartoshuk, professor of otolaryngology at the University of Florida, who finally solved this mystery, over sixty years later.

Bartoshuk, who wears her hair short and her eyeglasses big, became a taste researcher because of her father. "He had lung cancer when I was in college, and one of the things that bothered him the most was that his taste changed," Bartoshuk tells me over the phone in her strong, confident voice. "My aunt, his sister, made him a special kind of canned beef, hoping it would make him feel happier. But it tasted bad to him, metallic. It was bizarre. I realized only years later, when someone asked me why I study taste, that I have worked to solve this particular puzzle ever since."

But before she managed to find out how cancer changes our palates, Bartoshuk ended up solving the puzzle of Fox's PTC tasters. She started noticing that no matter whether she studied the strength of bitter, sour, or sweet taste perception, the same people always scored the highest. One day she asked her colleague, an anatomist, to come over to her lab to look at the tongues of the high scorers. "He was stunned," she recalls. "He said he had never seen tongues like that before." What her colleague noticed was that these supertasters (as Bartoshuk calls them) have more fungiform papillae and, as a result, a higher density of taste buds than do either medium tasters or nontasters.

If you want to find out whether or not you are a supertaster, Bartoshuk recommends a simple experiment: apply blue food coloring to the tip of your tongue and press a one-quarter-inch-diameter hole punch reinforcement label onto the tip. Then, looking in a mirror, count how many pink bumps are visible within the ring reinforcement; or ask a friend to count the bumps, using a magnifying glass. These are your fungiform papillae. The blue food coloring won't stain them, so they will look lighter in color than the rest of the tongue and can be seen quite easily. If you have thirty-five or more, you are likely to be a supertaster. And whether or not you are a supertaster has significant implications for the way you perceive taste. It may even affect how much you like the taste of meat.

My own counting experiment, just as I expected, proved I was a nontaster. I had only ten fungiform papillae inside the reinforcement label. Why did I suspect all along I was a nontaster? The signs were many. Nontasters, like myself, are not only much less sensitive to bitterness in foods than others but also more tolerant of oral pain, including spiciness. (Spiciness is not a taste, Bartoshuk tells me. It's a type of pain sensation.) Nontasters tend to like their coffee black and don't mind tannins in their red wine. They also tend to eat more vegetables of various kinds than do supertasters. The problem that supertasters have with veggies is that many of them contain bitter compounds called phytochemicals, which may act as natural pesticides, protecting the plant from parasites and predators. Beans, cabbage, brussels sprouts, zucchini, lettuce, grapefruit—the list of veggies and fruits loaded with bitter compounds is long. If you are a nontaster, you can happily enjoy such nutritious meat substitutes as dal or hummus, but if your taste buds are particularly plentiful, a vegetarian diet may be more of an uphill struggle for you. It doesn't mean, though, that you should just give up on bitter veggies altogether. Bitter phytochemicals such as phenolic compounds or flavonoids found in vegetables are actually very healthy. They lower the risk of cardiovascular disease and several types of cancer, may help cure some immune disorders, and help manage type 1 diabetes.

Yet although being a bitter-sensitive supertaster sounds like a perfect explanation for why some people may be slow to reach for healthy veggies and are hooked on meat, it's not so clear-cut. Some studies do show that supertasters eat less spinach and broccoli and like meats more than nontasters do, but other experiments report that instead of substituting bitter foods with animal protein, supertasters tend to gorge themselves on sweets. According to Bartoshuk, supertasters are more likely to be extreme about both sugar and meat, since in general they are more intense in their responses to food, strongly craving their favorite foods, while intensely disliking their least favorite foods.

Yet perceiving tastes is only a small part of what is going on in your mouth as you eat. You may notice creaminess, crunchiness, or sogginess—those are textures, detected by specialized neurons. In your slice of bacon, you may detect grassy notes or a hint of nuts and earth—those are aromas. As you chew, volatile compounds are released and drift up your throat toward your nose to be registered in a process called "mouth smelling." A large chunk of what we think of as meat's taste is actually its aroma.

The lack of strong aromas is one of the main reasons why we don't find raw meat very appealing. Even animals seem to agree: when scientists offer lab mice roasted ministeaks, for example, at first the animals are rather wary of this culinary invention. After a few bites, though, they go crazy for the cooked beef. In similar experiments, chimps, gorillas, and orangutans are clear about their preferences: roasting, grilling, and stewing make meat delicious. And one of the main reasons for that is the aforementioned Maillard reaction, the marriage between carbohydrates and amino acids in a slightly moist, hot environment (between 300 and 500 degrees Fahrenheit), which produces aromas so delightful they make us go weak in the knees.

Louis-Camille Maillard, the doctor who discovered this reaction back in 1912, looked like a hipster by modern standards, with his gelled hair, ultratiny round spectacles, and pointy mustache. Of course, in early twentieth-century France, where he lived and worked, this style was quite common.

Although Maillard's name is now practically synonymous with one of the most important reactions in food processing, the young physician wasn't originally interested in finding out why some foods taste delicious when they brown up. He was interested in kidney diseases. One day when he was heating sugars and amino acids together in a test tube, he noticed with surprise that the mixture turned brown at a lower temperature than expected. Maillard continued to study the reaction and in 1912 presented his findings to the French Academy of Sciences. But he failed to realize its gastronomical implications: he thought his discovery was mostly important for human physiology and for understanding coal and manure. Only later did he begin to see that the reaction he described was behind our love of some foods.

There are over one thousand substances responsible for the aromas of meats, and many are created in the Maillard reaction. Some smell fruity (γ-heptalactone), others musty (trimethyl-pyrazine), while still others may have the scent of nuts, mildew, smoke, marshmallows, or even crushed bugs (3-Octen-2-one). Although by themselves they may seem unappealing, taken together these substances comprise the mouth-watering aromas of meats. Protein hunger researcher Breslin jokes that even God appears to delight in the scents of the Maillard reaction. There are several passages in the Bible, he tells me, that mention how animal sacrifices as burnt offerings bring out "an aroma pleasing to the Lord." Why are we so attracted to these scents? One explanation is that in pre-refrigeration times meats got easily spoiled by bacteria, and the Maillard reaction was a way for us to notice that the food had been cooked and was safe to eat.

But there is a darker side to the Maillard reaction: it can produce acrylamide, a probable carcinogen, which forms when the amino acid asparagine (found in dairy products, beef, poultry, and eggs) is combined with glucose. Other products of the Maillard reaction have been associated with the development of diabetes, kidney problems, and cardiovascular disease. It seems that although the pleasing aromas of the Maillard reaction might have guided our ancestors toward safe and nutritious foods, in our age of refrigerators and antimicrobials, we should be less trusting of these tempting scents. Especially if we want to live past the age of thirty-five, the average life span of our Paleolithic, Maillard-reaction-loving ancestors.

Fat is another big piece in the puzzle of meat's delicious flavor. It is more energy dense than sugars and, as such, is highly desirable. For our ancestors' survival, it was vital that they identify and feast on foods loaded with fats, whenever they were available. After all, overindulging in very lean meat (the condition of most animals on the African savanna during the dry season) could lead to "rabbit starvation" and death.

It appears that human bodies just don't react well to diets in which over 35 percent of the calories come from protein—we need fat to dilute it. The lure of meat's fat is mostly in its aromas—all those sweet, charred, mouthwatering smells that waft out of the restaurant kitchens between Wharton Street and East Passyunk Avenue in south Philadelphia and which tempted me to buy a cheesesteak. But according to Breslin, instead of straying from the vegetarian path, I could have probably made my taste buds and brain happy with a serving of ice cream or something else that is fatty and tasty. "When you think you're craving meat, most likely what you're really craving is tasty fat. If you ate something that had almost no fat in it—a well-trimmed, lean piece of meat—it might not satisfy the craving. Hypothetically, though, a bowl of fatty ice cream could," he explains. When you cook a piece of pork or beef, it's not just the Maillard reaction that occurs. Fats start to oxidize, and even more delicious scents rush toward your nose. Fat is also where the most significant differences between the flavors of different species lie. Boiled or stewed beef smells mostly of an aldehyde called 12-methyltridecanal, which gives beef its unique tallowy and slightly sweet aroma. Another potent compound of the scent of beef is 2-methyl-3-furanthiol, which scientists describe as smelling sulphurous, sweet, and "vitamin." Chicken, meanwhile, smells of trans-2-trans-4-decadienal, which at low concentrations has an aroma of oranges or grapefruits.

Yet it's not just the smell of fat that keeps us hooked on meat. It's the texture, too: creaminess, juiciness, and crunchiness are all sensations that tell us that the meat is loaded with fats. Neural-imaging studies show that specialized fat-sensitive neurons in the brain respond to the lubricity (slipperiness) of fat in our mouths, which registers as a pleasurable experience. What's more, over the last ten years evidence has been mounting that we can detect the taste of fat in our mouths, in a similar way that we taste salty or sweet—which would make fat the sixth basic taste. This detector system informs our brains that fat is present in the food and can prepare the body to digest it, as well as reward us with pleasure for eating it. Painting your tongue blue can also give you an insight into how well you detect fat in foods. The more mushroom-like fungiform papillae you have, the better your ability to detect fat in some foods, studies show. For example, supertasters may excel at telling whole milk from a low-fat one. But nontasters have their own ways of responding favorably to fat: experiments show that nontasters prefer their food high in fat, since they are so inept at perceiving it at low concentrations.

But not all fat in meat tastes equally good. Imagine you got magically transported back in time into the depths of the Mesozoic era, into a towering forest full of ferns and gingko plants. You conveniently have brought with you an impressively big shotgun. You are starving. You need to find some food ASAP. Yet there are no familiar species of animals that could give you an idea of what to hunt. As you look around, you notice a few dinosaurs of different species grazing or walking around. Assuming they would all require similar skill and effort to hunt, which one should you kill? Which of the dinosaurs would make for the best steak? A *Tyrannosaurus rex*? One of the long-necked sauropods—the largest animals to ever walk the earth? Or maybe an ostrich-like ornithomimid—one of the fastest runners of the dinosaurs' world?

The secret to the tastiness of an animal's meat lies to a large extent in its diet, which translates into the composition and flavor of its fat. The majority of people around the world prefer meat that comes from herbivorous animals, such as cows, sheep, or deer, since a carnivore's diet adds a gamey flavor to the animal's fat that is not very appetizing. That's one of the reasons why lion burgers and cat steaks are not exactly popular and also why the meat of a *T. rex* likely wouldn't be very tasty. You'd be better off hunting the ornithomimid. Because of its very active lifestyle (ornithomimids ran a lot), that particular dinosaur's meat would be composed of slow-twitch muscle fibers, which means it would be red, a bit like beef. The high activity levels of ornithomimids would also mean that their flesh would be quite rich in umami—the fifth basic taste, which is the last clue to why humans love the taste of meat.

It took a Japanese chemist named Kikunae Ikeda just a little over a year to figure out what tomatoes, meat, and kombu kelp have in common. In Kyoto, where Ikeda was born in 1864, kombu was commonly used to make a broth called dashi, which in turn served as the base for miso soups and noodle dishes. Ikeda, a slim, frail-looking man, noticed that dashi has a very distinct taste, quite different from salty, sweet, bitter, or sour. He also noticed that whenever his wife cooked soups based on dashi, they were particularly delicious. One day in 1907, Ikeda (who by then was a professor of chemistry at Tokyo Imperial University) took a huge evaporating dish, filled it with water and eighty-four pounds of dried kombu, and let it all simmer. From the broth he obtained, he managed to extract one ounce of monosodium glutamate, which is the sodium salt of an amino acid glutamate. That was it. Ikeda was convinced that monosodium glutamate gave the specific taste that was behind the deliciousness of not just dashi but also of tomatoes, cheese, and meat. He called this new taste umami—Japanese for "delicious."

Unlike Maillard, Ikeda knew right away that what he had discovered not only was fundamental for science but also had practical applications for the food industry. He quickly obtained a patent for a seasoning based on monosodium glutamate and contracted Saburosuke Suzuki, a businessman who worked in iodine production. Suzuki was impressed with Ikeda's discovery and decided to invest in it. Soon Suzuki Pharmaceutical Company started to manufacture and market monosodium glutamate under the name Ajinomoto, Japanese for "quintessence of flavor." Today, Ajinomoto Company is a major Japanese corporation and operates in twenty-six countries. It sells 40 percent of the world's aspartame sweetener and is the largest producer of MSG (monosodium glutamate) seasoning on the planet. If you go to your nearest well-stocked grocery store to buy MSG, it will most likely come in the characteristic red Ajinomoto packaging.

From the beginning, though, Western scientists were skeptical of Ikeda's discovery. So skeptical that it took almost a century for "deliciousness" to become widely accepted as the fifth basic taste. Most Western scientists believed umami to be no more than a combination of the other four tastes and claimed that re-creating umami is just a matter of finding the exact proportions of salty, sweet, sour, and bitter. Yet no one ever managed to achieve that—and not for want of trying. It was only in 2000 that umami finally got its big break. That year three American scientists located umami receptors on human tongues, a discovery that was soon confirmed by other researchers. Umami officially entered the pantheon of basic tastes.

In nature, three substances are responsible for umami taste: amino acid glutamate (which is behind the deliciousness of Parmesan cheese, soy sauce, sun-dried tomatoes, and cured meats) and two nucleotides—inosinate (IMP), found in meat and fish, and guanylate (GMP), present mostly in mushrooms. When one or both of these nucleotides are combined with glutamate, the umami taste of a dish is magnified to as much as eight times that of glutamate alone. This synergy is why pepperoni pizza and Pat's Philly cheesesteaks are so mouthwateringly delicious—it's the glutamate from cheese, plus IMP-rich meat and GMP from mushrooms (as long as you order your cheesesteak "wit" mushrooms, of course). Meat is a particularly good example of umami's mouthwatering synergy: because it contains the umami compounds glutamate and inosinate, it produces a strong and long-lasting "delicious" sensation.

All this is no secret to the best chefs—some of whom call dishes that profit from the umami synergy "u-bombs" and who eagerly use these substances in their cooking. Adam Fleischman, the chef of the famed Umami Burger restaurant in Los Angeles, named his signature umami burger "umami × 6." It maximizes the delicious umami sensation by combining beef with grilled shiitake mushrooms, roasted tomato, caramelized onion, homemade ketchup, and pan-fried Parmesan. Another celebrated chef, Sat Bains, who heads the kitchen of a Michelin-starred restaurant in Nottingham, UK, creates a delicious "u-bomb" by brining beef overnight in a kombu solution, to benefit from the synergy of the inosinate in meat and the glutamate in the seaweed.

Humans love umami. We learn its taste even in utero (the amniotic fluid contains glutamate) and then at mother's breast, as human milk is particularly rich in umami, much richer than that of most other mammals. Yet many of us in the West have trouble telling umami taste as easily as we can pinpoint sweet or salty. You won't hear many Americans or

British say: "Oh, that soup is not umami enough." So what does umami actually taste like? To find out, try an experiment recommended to me by Beauchamp during my visit at Monell. Take two cups of soup. To one of them add a sprinkle of MSG, which you can buy in most well-stocked grocery stores. Take a sip of each soup and compare—the difference should be obvious. When I did this experiment, I knew right away which soup had MSG in it. It tasted much fuller than the other one, more savory and round. It tasted "meaty," as umami is often described, even though there was no meat in the soup. I could feel the umami taste not only on my tongue but all around my mouth—on my palate, on the insides of my cheeks. Did it taste better? You bet.

If the soup experiment doesn't help you pinpoint umami, you may be umami blind. About 3.5 percent of the population are unable to detect the umami taste of monosodium glutamate. Meanwhile, a study of twins showed that our liking of protein foods such as meat and fish is the most heritable of all food preferences. The reason may lie in the way we respond to umami and that some of us are genetically better at discriminating this particular taste. Other experiments point in the same direction: people who are the most sensitive to the taste of MSG report a greater liking and preference for protein-rich foods, like meats. Could that mean that the 3.5 percent of us who are insensitive to umami are more likely to become vegetarians? And is it just a mere coincidence that in many Western countries the percentage of vegetarians oscillates around 3.5 percent? Maybe that's all it is—a coincidence. Maybe not. So far there have been no studies that examine how insensitivity to MSG may translate into choosing a vegetarian diet. If you know any academics doing research on nutrition, ask them. Maybe that question will spike their interest enough to investigate.

There are some vegetarians out there, though, who stick to their diets precisely because of their inability to taste umami. Although giant pandas (these are the vegetarians I'm talking about) officially belong to the order Carnivora, along with such avid meat eaters as lions and wolves, they are basically vegetarians. Their diet is 99 percent composed of everything bamboo: bamboo leaves, shoots, and stems. In one year a single panda devours over ten thousand pounds of bamboo. The remaining 1 percent of the giant panda's diet comes from such bamboo seasonings as grasses, bulbs, and insects. The thing about giant pandas is that although they have the short digestive system of a carnivore, their umami taste receptor gene is not functional, and as a result, giant pandas can't taste umami. The loss of the umami perception may explain why giant pandas are not interested in meat anymore. Over generations, their decreased reliance on meat could have resulted in their loss of umami taste receptors, which helped to keep the animals hooked on bamboo. Since pandas couldn't taste umami anymore, they stopped liking meat and eating it.

Why does umami taste so delicious to us? Why do we seek out foods that, just like cooked meat, are brimming with umami? Most scientists believe that umami signals the presence of proteins in foods and helps us choose good sources of that nutrient. Meat, of course, is rich in proteins. But umami is not just about proteins—it could also mean that fewer harmful bacteria are present. Cooking and aging break down large protein molecules in meats and release glutamate, making foods more umami. Umami taste, just like the products of the Maillard reaction, may signal that the meat has been cooked—and is safer to eat. But there is still a lot of mystery surrounding umami. We don't know why human milk is so loaded with it. We are not sure which genes can make humans umami blind. We don't know the purpose of umami in tomatoes, since they are not exactly brimming with proteins. "What we do know," Beauchamp tells me in his cluttered Indiana Jones–like office, "and that's something I'm *very* convinced about, is that umami is a particularly potent enhancer of pleasure in certain foods, including meat." And if you can't taste umami, bacon, burgers, and cheesesteaks won't be as yummy to you as they are to others.

Are we then doomed to crave beef and chicken and pork, with their delicious umami, mouthwatering fat, and aromatic products of the Maillard reaction? Are we going to stay hooked on meat no matter how bad all these cheesesteaks and burgers may be for our arteries and for the planet? After all, even many vegetarians are not immune to the lure of meat's flavors. According to one survey, 60 percent of vegetarians admitted to having eaten meat within the past twenty-four hours.

But knowing what precisely makes meat so delicious and enticing—which exact aromas and umami substances and which fat textures—can help in the future to create the perfect meat substitute, the vegetarian Holy Grail. There is still a lot of work left for scientists to do, of course. Out of at least a thousand substances that create meat aromas, we know only a few. We don't understand the perception of fats very well. We don't know exactly how umami works and why some people are blind to it. There are many studies and experiments to be done. It requires time.

Years of evolution have taught us to seek out cooked meat for the nutrition it offers. We are genetically programmed to respond favorably to umami, which signals the presence of proteins, and to the aroma of cooked meat (the Maillard reaction), which signals that the food is safe to eat. Of course, it does not mean we should just follow our taste buds blindly and munch beef burgers like pandas munch bamboo. The meats of today are not the same as the meats of the past. Our lifestyles have changed. The way we rear animals has changed and so have our bodies—even if our taste genes lag behind. Who knows: maybe they will also adjust one day and make our palettes better suited to the modern world. But we don't have to wait for a panda-like mutation in our taste genes to occur. We can use the knowledge about what makes meat delicious to replicate these sensations with other foods that are better for our health and for the environment. Those who want to cut down on meat can try a few simple tricks. For your Maillard reaction fix, choose freshly baked breads, crunchy toasts, and roasted veggies. To make up for fats in meat, go for avocados, cheese, and nuts. As for umami, try tofu cooked with soy sauce, a little peanut butter, and mushrooms. Even if these foods don't exactly taste like grilled steak, they can still make your taste buds very happy.

Yet such culinary tricks won't delight everyone—not the meat industry, for one. After all, the industry works hard to ensure that we will like the taste of meat more than the taste of its substitutes: animals are bred, fed, and killed in ways that win the battle for our palates and our wallets. Meat producers massage cows with sake, feed green tea to chickens, castrate baby pigs (with no anesthesia), and inject meat with brine—all the while struggling to keep a perfect balance between the taste of meat and its cost so that we keep coming back for more and more and more.

5

Why Would Abramovich Taste Good?

"Come and get your beer, boys!" Ifor Humphreys calls, as he splashes a bucketful of award-winning ale into a wooden trough. Seven black steers lazily raise their heads, their mouths full of hay. "Not too thirsty today?" Humphreys asks, and as if in response, two of the cattle turn toward the trough and dip their long, black tongues into the frothy brew. It's beer time at the farm. Shortly after, it's massage time. Humphreys pulls on his rain boots, picks up a currycomb, and steps between the steers. He begins to brush one of the massive black backs, just the way you would groom a horse. Swipe, swipe, circle, circle, swipe. Humphreys doesn't massage his beef cattle every day, but he does it often enough for their coats to shine. Although the steer doesn't exactly sigh in pleasure, he does seem unusually content, relaxed. The beer and massage, a Kobe-style treatment, must be working.

If the name Ifor Humphreys doesn't sound very Japanese to you, that's because it isn't. The "Kobe" farm I'm visiting is not in Kobe, Japan, but in Abermule, nestled in the hilly greenness of Wales. Yet Humphreys's beef has much in common with the meat from Kobe. Humphreys puts in many hours a day to produce beef that tastes better than the regular supermarket cuts, and judging from his satisfied and growing customer base, he is succeeding.

It comes as no surprise that most modern food is carefully engineered. Few consumers would be surprised to learn that Oreo cookies and Cheez Whiz have been painstakingly designed in laboratories to be as pleasing to our taste buds as possible. But fresh meat producers work hard on their merchandise, too. If you are a meat company, and you want people to buy lots of meat, you have to make sure that the product you are selling is tender, juicy, and full of flavors. It's not an easy task. Many large meat producers have ultramodern research facilities, where a legion of meat scientists design animal feed, study breed genetics, and devise new ways to package the products so that they stay fresh and appetizing for as long as possible. Their goal is to ensure that Americans—and other nations—stay hooked on meat.

Humphreys is not a scientist—"I'm just a farmer," he says—but he knows a fair bit about how to make beef taste good. And just like many meat scientists, he believes that the breed of the animal is of particular importance: some breeds of cattle, poultry, or swine simply taste better than others. Take Berkshire pigs. According to a legend, over three hundred years ago, when the army of Oliver Cromwell was stationed in the shire of Berks, just west of London, the soldiers came across a remarkable breed of hogs. The animals were huge and produced hams and bacon so delicious their fame soon reached the ears of the British monarchs. A herd of Berkshire pigs was installed by the Windsor Castle to provide the fatty and amazingly tender pork for the royal tables. Years later, the breed benefited from a bit of Chinese and Siamese blood, which made the pigs even more efficient at plumping up. When the Japanese from Kagoshima Prefecture got hold of Berkshires, they called them Kurobuta (black pigs) and worked hard at breeding to make the meat even better. Today Berkshires are famed worldwide and sometimes even called the Kobe of pork. The official website of the Kagoshima Prefecture boasts that Kurobuta pigs have a "lip-smacking flavour" and meat that is light and crisp, soft and tender.

The tenderness of meat is one of the key features that the industry works to improve. Meat scientists say that tenderness is all about resistance to tooth pressure. If meat is not tender, at some point in the chewing process you will be ready to swallow but at the same time you will feel like you should chew a bit more. And, most likely, you will end up forcing a sinewy blob down your throat. On the other hand, meat that is tender will be soft on the tongue and leave no coarse strands between the teeth. Not everyone agrees that increased tenderness is better. In some cultures—in parts of Africa, for example—people actually like their meat to be chewy. In the West, though, the preference is for soft and tender, especially among women.

The science of meat's tenderness is quite complicated, yet there are some predictable qualities that can push a steak or a chop from being soft to resembling a piece of dried-out rubber. To be tender, a muscle should contain low quantities of collagen—a type of protein that is the main component of connective tissue and that is abundant in tendons, skin, bones, and blood vessels. Some muscles have tons of collagen, making them tough, others very little. If you like your meat soft under your teeth, here's a tip: avoid cuts that come from powerful muscles that the animal used a lot. If a muscle is exercised frequently, like the ones found in legs, it needs a lot of support in the form of connective tissue made of collagen. Most tender meat comes from muscles designed for structural support rather than for movement, such as the ones running along the spine.

There is one particular thing about the Kurobuta breed of pigs that makes their meat notably tender: marbling. Marbling occurs when fat is deposited between bundles of muscle fibers—that is, inside the muscle itself. If you take a piece of meat that has very little marbling, it will appear almost uniformly red. A well-marbled meat, on the other hand, will have so many white intrusions of fat that there won't be much space left for the red. Marbling not only makes meat more tender (fat is softer to bite through than muscle) but also more juicy. Such fat gets released from the meat when you chew it, stimulating the flow of saliva and providing that "melt in your mouth" sensation.

Marbling is also the main reason why the word *Kobe* is for many practically synonymous with "deliciousness." Traditionally produced among the green mountains of Hyogo Prefecture in Japan, Kobe is so renowned and craved that its price may soar up to $400 a pound. When Mark Schatzker, a *Slate* columnist who has traveled the world in search of the most perfect steak, tried his first cut of Kobe, he described it as being "smoother than hot buttered silk." Kobe beef is so good mostly because the black Wagyu breed of cattle from which it comes produces extremely marbled meat. Some cuts of the A5 beef (the highest grading of marbling) are so fatty they look like a piece of lard dotted with a touch of ketchup. Humphreys is one of the very few farmers outside Japan who owns the famous black Wagyu breed. "In all of UK there is probably ten of us," he tells me. If you are not in Japan, it's not easy to get hold of Wagyu genetics to start your own herd. Seven years ago, when Humphreys decided to do something more "exciting" and try his hand at producing Kobe-style beef, he had to import a Wagyu embryo in a flask of liquid nitrogen all the way from Australia and implant it into a surrogate mother. He got a pure-blood Wagyu cow this way and a bull whom he named Abramovich (after the Russian investor, so that it sounded "contemporary and prestigious"). Fast-forward seven years, and now Humphreys has over twenty head of cattle, although some of them have a touch of Angus blood mixed in. That's Humphreys's little "Kobe" herd.

In Japan, no cow is born a future Kobe beef, though. Only a chosen few, those that meet several specific criteria, are "entitled to be called 'Kobe beef,'" in the words of the Kobe Beef Marketing & Distribution Promotion Association. Either a pure-blood bull or a virgin Wagyu cow has to be raised and killed in Hyogo Prefecture. Its meat has to have tons of marbling and a firm texture. Each year as few as three thousand such cattle get certified—after death, of course—as Kobe beef. Very little of it makes it outside Japan. If you think you ate Kobe beef in the US before 2012, you can be certain you got scammed (unless someone smuggled it for you in a suitcase). The first cut of true Kobe arrived officially in North America in November 2012, imported by the Fremont Beef Company. In all of 2013, just 3,636 pounds was imported to the US and none to Canada, Australia, the UK, or any other EU country. Official imports to Europe began in mid-2014, with shipments to Germany and the Netherlands. In the States, you can eat a real Kobe steak at only a few restaurants, such as one at the Wynn resort in Las Vegas or 212 Steakhouse on East Fifty-Third Street in New York. Most of the other

"Kobe" is not the real thing. Some of it may be from Wagyu cows raised outside Japan using traditional techniques, Humphreys style. The rest is regular beef, just overpriced.

Although Kobe is hard to come by in the US, there is plenty of Certified Angus Beef to please those who crave fatty steaks. Since cattle of the Angus-Aberdeen breed pack on fat earlier in their lives than most breeds, their meat is usually more marbled and tender than that of other breeds. The Certified Angus Beef, meanwhile, is a *brand* of beef. There are ten specifications that a cow's carcass has to fulfill to be recognized as Certified Angus Beef—all in the hope that the meat will be so good that customers will keep wanting more. It has to have abundant marbling, a well-shaped rib eye, superior color, and "no neck hump exceeding 2 inches." The argument behind this last requirement is that a neck hump is characteristic of Brahman cattle, famous for being the typical "holy cow" of India and for its particularly tough meat (which could well be one reason why Indians didn't develop much taste for beef).

If you want to keep people buying the meat you produce, it's not just the breed that counts, though. What you feed your animals is also important. The grain or grass they consume has a large impact on flavor. Diet was the reason, for example, why pork in colonial America wasn't as good as the pork of the early twentieth century, nor even like what we have available to us today. In those days pigs ran freely in the forests, rooting for acorns and nuts. This made for happy pigs but poor meat—a diet of acorns makes pork soft, oily, and prone to rancidity. As a result, the original American pork couldn't compare with modern, corn-fed industrial pigs, far from happy animals but better meat.

Neither Kobe nor certified Angus cattle are by definition grass fed, yet grass-fed beef has been on the rise in recent years as a natural, healthier option. But if we were to ask whether such beef actually tastes better, the answer would depend on the person asked. A majority of Americans don't actually like the taste of grass-fed beef: it's too strong for them, too gamey, and some have even described it as "fishy," a flavor that comes from a high amount of omega-3 fatty acids found in such meat—the same ones that are abundant in salmon or mackerel.

Generally, the argument goes, you like what you grew up eating. If the beef of your childhood was corn fed (as in the US), you will enjoy corn-fed beef as an adult. If you ate barley-fed cattle (as in Canada), you will like barley fed now. One problem American consumers have with grass-fed beef is that it tends to have yellowish fat. There is nothing wrong with yellow fat—the color comes from the carotene animals get from eating green plants. But Westerners don't like the fat in their meat to be yellow. They prefer it to be snowy white, standing out in contrast to the bright red of the muscle.

You may have heard that the secret of Kobe beef lies in the way the cows are pampered. Stories abound of cows whose backs are massaged with sake, their thirst slaked with beer, and their sense of pleasure stimulated by Mozart's piano concertos. But it's all a myth—at least as genuine Japanese Kobe goes. According to the official website of the Kobe Beef Marketing & Distribution Promotion Association: "There have been almost no cases of cows being raised on beer," and "massage itself neither softens meat nor increases the amount of marbling." Humphreys knows that the way he treats his cows is not standard in Japan, but he does it anyway. Why? "It's a good story," he says, and then adds, with a smile: "And the cows like it."

Although massaging cattle hasn't been shown to improve the taste of their meat, we've long known that, on the other end of the spectrum, stressed, anxious animals all too often equal bad meat. If you find that your sautéed pork medallions are devoid of flavor, you may have just eaten a piece of an animal that has endured a life that couldn't be further from the lush hills of Kobe or the rolling fields of Wales. It might have grown up in such filthy, crowded conditions that, out of stress, it compulsively gnawed on its cage bars. It might have been roughly handled, even beaten, and kept without adequate water. It might have been hungry. Because of overbreeding, it might have been too crippled to walk. In its final trip to the slaughterhouse, it might have been transported for days in the soaring heat of a truck, pressed against companions that had already died from the strain. Under such prolonged stress, muscle fibers become packed too tightly, like yarn in a sweater washed in hot water. The result is toughness, dryness, and bad taste. This condition, known as DFD meat (dark, firm, dry), affects about 10 percent of the pork in both the US and the UK and about 15 percent in Australia. In other words, far more farm pigs, cows, and chickens experience this kind of stress, which influences the taste of their meat, than enjoy the luxury treatment afforded to a tiny portion of animals. And it's surely not the only industry practice that takes its toll on taste (not to mention the well-being of animals). To learn more about how meat producers end up influencing their own products—and potentially put up roadblocks to keeping consumers hooked on meat—I decided to seek out the experts and headed to State College, Pennsylvania.

I'm wearing a blue hard hat and a white overcoat that is far too big for me. The corridor in which I'm standing is spartan—the bright overhead lights giving it a slightly unreal, washed-out appearance. Here at the Penn State University Meats Lab, a sixteen-thousand-square-foot facility in State College, designed for research of animal slaughter and processing, I've come to learn how, and how *not*, to make meat taste good. Edward Mills, professor of meat science, is a big man who looks perfectly comfortable in a workman's outfit. Once an aspiring vet, he grew up on a farm, where they regularly slaughtered hogs and cattle, so his work researching the death of livestock doesn't bother him much at all.

As we tour the lab's facilities, Mills offers me a sniff of one small vial that contains a substance that plays an important role in how good meat tastes. For Mills, the substance has no scent, but other people with a different genetic makeup are able to detect the odor. Unfortunately, it seems I'm one of them. The moment I place the vial under my nose, I feel as if someone punched me in the face. The stench is unbelievable. Imagine a mix of rotten rat carcasses with sweaty unwashed feet, and then multiply that by a hundred. The substance is called androstenone, and it's a compound produced in the testicles of male pigs. Androstenone can make pork smell disgusting when cooked, and so for the meat industry, it spells trouble.

The simplest solution to the problem of androstenone is to castrate the pigs. But the customers not only want their pork free of dead rat stench, they also want it to be lean. Intact boars don't get as fat as castrated ones do, so their meat is better for your arteries—and worse for your sense of smell. If boars could choose, they would probably vote for androstenone-tainted pork. Castration is usually done on fully conscious animals, without anesthesia or painkillers, by making incisions with a scalpel and squeezing the testicles out. It takes a long time. That the animals suffer is obvious from their cries (for those with steel nerves there are videos of the procedure on the Internet).

It seems, though, that you can get used to the stench of androstenone (luckily in meat it's much more diluted than the concentrated form I experienced). In a few countries boars are not castrated, and their meat is more likely to be tainted with the compound—and yet, people still eat it. But if you are an American who grew up consuming pork from castrated pigs, you may be in for a culinary surprise in places such as Ireland, Spain, or the UK, where male pigs are usually—and happily—left untouched.

With the vial of androstenone safely stowed away, we go upstairs to visit the kill floor. Thankfully, there are no animals there today, and I am able to walk down the ramp, just as a cow or a pig would, between the bare, gray walls. At the end of the narrow, winding path, in a blood-stained room that smells of metal, is an electric stunner, a gun-like machine that renders an animal unconscious. Its electrodes are applied to the animal's forehead, electrical current passes through the brain, and the animal collapses, its head extended. The breathing stops, but after fifteen to twenty seconds, the legs start to kick. Just then a knife is stuck in the animal's throat to severe the carotid arteries and jugular veins. If it's done correctly, brain death will occur about twenty seconds later. The animal bleeds out. Every Tuesday between twelve and twenty animals are slaughtered at Penn State—a tiny amount compared to big "processing plants" (meat industry lingo), which can "harvest" (an industry euphemism for "kill") about 400 cows, 1,000 hogs, or 46,000 chickens *per hour*. Every day in the US, 24 million farm animals walk down ramps similar to the one I walked at Penn State, or are put on conveyor belts that air-lift them to the killing floor. That's as much as the human population of New York, Los Angeles, Chicago, Houston,

Philadelphia, Phoenix, San Antonio, San Diego, and Dallas *combined*. About 9 billion animals per year in the US alone.

Mills's lab has confirmed that the last hours and minutes of an animal's life are particularly important in determining the taste of the meat on your plate. The more it suffers, the less likely your pork chop or steak will be any good. "If the animal is severely stressed, the meat may be unusually dark or pale in color and may have an off flavor. The handling right before the slaughter is also really important for meat tenderness," Mills tells me, as we are standing among the various hooks, chains, and knives—the "harvesting instruments." What the meat industry is particularly worried about is the so-called PSE meat, short for pale, soft, and exudative (oozing fluid), a condition that can mostly be seen in pork but can also occur in other types of meat, like turkey and chicken. Next time you cook a piece of pork that is pale pink and tastes quite bad, you may well assume it came from a pig that suffered acute stress just before it got killed. In its final moment, it was likely panting with its mouth wide open, squealing and trembling, its skin covered in blotches. In the US, 16 percent of all pork is PSE. In the UK, up to one-quarter is, and in Australia, even as much as 32 percent. Stressed by the restraining conveyors, which immobilize them and carry them forward to the slaughter point, poked with electric prodders, and killed alive after an inefficient stun, no pig's meat will end up tasting good, regardless of its breed. Once "they die piece by piece," in the words of a slaughterhouse worker in an interview with the *Washington Post*, stress hormones like adrenaline will flood the body, and the temperature of the animal will rise above normal. Setting aside the matter of cruelty, this poses a real problem for meat quality.

Making matters worse, after death the muscles acidify. This is essentially the same process that happens when you run really fast for a short distance: your muscles work without adequate oxygen, causing the formation of lactic acid. Once you stop running, your liver will clear the lactic acid from your body. You may be sore the next day, but you will otherwise be fine. In a dead farm animal whose liver is not working anymore, the lactic acid will lower the pH of the muscles far below normal. Such acidification, if combined with elevated temperature (due to acute stress), will cause the meat proteins to lose their normal structure—the same thing that happens in the body of a very sick person when his or her fever rises above 107.6 degrees Fahrenheit. Denaturation of proteins makes them less efficient at binding water. Instead of staying inside the meat and making it moist and juicy once cooked, the water drips out. It also takes with it the water-soluble red pigment, myoglobin, so that the muscle will lose its appetizing color. In meat-science lingo, such meat has low WHC—water-holding capacity. You probably have seen quite a few examples of that problem on the supermarket shelves: these are the packages of meat that have accumulated a pool of bloodish slop at the bottom of the Styrofoam tray. That slop, called purge or weep by the industry, is a solution of water and proteins—and a sign that the meat may have come from an animal that suffered even more than average. To avoid its unappetizing sight, meat producers add absorbent pads at the bottom of containers to soak up the liquid. It may work to make the product look better, but if you cook a meat with low WHC, it will likely end up tough and dry.

If an animal spent its life chronically stressed and experienced acute stress just before death, the result will be decidedly unappetizing. Some of its muscles will be pale and drippy, while others will turn dark and dry. As a result, one would quite reasonably conclude that, not just morally, but in the interest of producing a high-quality product, it would be best if animals suffered no stress in their lives at all and if the killing process was swift and painless. But the high incidence of PSE and DFD meat reveals that there are other dynamics driving meat producers. Though there are many ways to lessen the stress of farm animals, they are often just not cost efficient. There would be less PSE meat out there if the speeds at which animals are stunned and killed were slowed down, allowing more time for precision. But this would raise the costs, and making meat is a business, just like making furniture or mascara. The numbers have to add up.

Meat producers constantly struggle to balance quantity against quality. Keeping consumers hungry for meat means playing a game of tug-of-war between price and taste: the meat has to be flavorful, juicy, and tender, but it can't be too expensive. Humphreys's cattle live as stress-free lives as meat cattle can, and the massages and the beer may help keep them even calmer, making DFD profoundly unlikely. But their meat is expensive, at thirty-seven dollars for a single six-ounce fillet steak. Not everyone is willing to pay the price for stress-free meat. And so the struggle—cost versus quality—continues.

As Mills and I stand within the sterile whiteness of the coolers, only the reddish carcasses of pigs hanging from the ceiling add a splash of color to the room. They look naked, these animals—naked and cold. The chill is creeping into my bones, and the sweet scent of dead muscles, of fat and blood, is making me dizzy. I want to move on, but Mills has a story to tell me, and the best place to tell this story is here, in the coolers.

If you end up with a piece of pork or beef on your plate that is unpleasantly tough, Mills tells me, it doesn't necessarily mean the animal suffered excessive stress just before death (it was most likely stressed, mind you, just maybe not *unbearably* stressed). The problem may have arrived postmortem. A few years ago Mills was asked by a pork producer to solve his meat-tenderness problem. His customers seemed to be losing the taste for pork. According to Mills, the meat was truly horrible. Yet it didn't take him long to figure out what the problem was: cold shortening, as bad as it gets. Cold shortening happens when you chill the carcass too fast and too much: the muscles contract considerably more than they usually would after death, which leads to lack of tenderness after cooking. "Imagine taking a bundle of rubber bands that you are somehow able to bite through—that's the texture of cold-shortened meat," Mills told me. The Pennsylvania producer was blast chilling the dead pigs with ultracold air to speed up the lines and get a ready product in eighteen hours instead of the usual twenty-four. Mills told him to raise the temperature in the chillers—and voilà—problem solved. The loin chops became tender again.

Sometimes meat producers try to chill the carcasses faster than advisable to make up for the fact that their meat comes from acutely overstressed, scared animals. The reasoning here is that if they manage to get the adrenaline-induced heat out of the animal's body before the muscles acidify, the meat will be just fine. But that often backfires, leading to cold shortening instead, and the consumer still ends up with tough, chewy steaks or chops. There is just no escape, it seems. Miserable, suffering animals generally equal bad meat.

There are exceptions, however. Some animal suffering actually results in better-tasting meat. For one, there is the androstenone-free pork from male pigs that endured castration without pain relief. But the classic example is pale veal. This doesn't come from any specific breed of cattle; it's all in the upbringing. To produce pale veal, you take young animals from their mothers soon after birth, stuff them into tiny cages that don't allow for any movement, chain them at the neck, and feed them milk formula that is so deficient in iron they will become anemic and extremely weak. Such calves can't move around, can't stretch their legs, and if released, may not even be able to walk anymore, not even to the slaughterhouse. But meat produced this way will be tender and delicate, with creamy white fat. Why is the meat so good? The paleness is due to the animal's anemia—their red blood cells don't contain enough hemoglobin, a protein responsible for carrying oxygen throughout the body and for the red color of the blood. It is tender because the animals never move, so the muscles are low in collagen. The veal crates designed to keep baby cows immobile were banned in the EU in 2007. The American Veal Association is "encouraging" farmers to adopt group housing methods by 2017, but as long as customers want their veal pale and tender, and the law doesn't say no, the veal crates will likely stay.

For years Temple Grandin has been working with the meat industry to improve the ways in which animals are treated—and which would also improve the taste of meat. An animal-wellness advocate born with autism, she was listed by *Time* magazine in 2010 among the hundred most influential people in the world. That same year HBO released a movie about her, which was nominated for fifteen Emmy awards and won seven.

Grandin has taken on many animal welfare issues: she has opposed electric prods, overcrowded pens, overloaded trucks, and rough handling. Now one of her pet projects involves opposing the use of beta-adrenergic agonists, a type of hormone-like drug. If fed to cattle, Merck's Zilmax or Eli Lilly's Optaflexx makes the animals get much more muscular. Add some Optaflexx to the feed just before slaughter, and your cow can gain an additional twenty-two pounds. But the problem with beta-adrenergic agonists is that although they boost *quantity* of meat, they may destroy *quality*. Meat from animals fed these drugs is often dry, dull in color, and far from tender. Again, chances are good that if you buy a piece of meat in the US and, after cooking it, discover it is not as succulent as you expected, it came from an animal raised with the help of beta-adrenergic agonists. Seventy percent of US cattle are given such drugs to promote growth. These animals often truly suffer. "Hot weather makes it worse," Grandin once told me. "I've seen groups of cattle in slaughterhouses in hot weather. They acted like their muscles were stiff; they didn't want to move. In rare extreme situations, their feet may fall off." Such cows, whose feet—as one animal science professor once described—are "basically coming apart," would not make a mouthwatering steak, that's for sure.

Yet the meat industry doesn't want to lose your taste buds to soy steaks and meatless meatballs—they want you to find their products delicious, tender, and juicy. At the same time, they want to grow as much muscle as fast and cheap as possible, and then "harvest" it as fast and cheap as possible. But what if because of all this rushing and pushing and growth promoting, the meat ends up being tough and dry? Luckily (for the meat producers, at least), not everything is lost, even if the carcass has already been dressed and chilled and is ready to be packed onto Styrofoam trays. What you need at this point is a needle and a chemical solution, a solution you can inject into the meat to make it tender once again, or at least tender *enough*. To achieve this, you can insert a solution of salt, phosphate, and lactate into the meat, which will improve both tenderness and juiciness. You can also infuse it with special enzymes that break down proteins. There are many chemicals that get injected into fresh meat in North America and Australia. According to the beef industry, "Enhancement generally ranges from 6% to 12% of initial weight." It may include substances such as hexametaphosphate, sodium tripolyphosphate, tetrasodium pyrophosphate, sodium lactate, and calcium lactate. They not only make the meat more tender, but also boost juiciness and add "meaty" flavors.

If injecting the meat with brines doesn't work, you can always mince or flake the muscle into a pulp and, with the help of a glue, shape a new, better steak or a chop out of it. The glue can be sodium alginate, a gel made from brown seaweeds, or transglutaminase, a bacteria-synthesized white powder you can dust over pieces of meat to stick them together into a "steak." (If you want to trick your dinner guests into thinking that you are serving expensive cuts and not cheap mince, you can buy some under the brand name Activa from www.amazon.com.) Some processes of this so-called meat restructuring are so efficient that you may not even be able to tell that the "steak" you are eating is not really a steak. You are most likely to encounter such products in chain restaurants and cafeterias and in frozen-meat cases in supermarkets. Restructured meats are usually rib- or steak-shaped and often breaded or sold as part of ready-made dinners. A survey in the UK showed that some such "meats" contain only 55 percent actual meat (the rest is soy and other compounds), so if you eat them, you could say you are already halfway to being vegetarian—and switching to 100 percent soy burgers wouldn't be much of a culinary shock to you.

People want the meat on their plates to be tender, juicy, and bursting with flavors. Meat producers want this, too. For them, whether or not we keep buying meat is a matter of survival. And so they breed the animals for marbling, massage them with sake, keep them imprisoned in tiny cages, shock their bodies with electric currents, and inject them with brines. Every day, they look for new ways to make the meat tastier and keep us craving for more. They research feeding green tea to turkeys, offering garlic to lambs to improve meat quality, and using gas to kill chickens to make the meat less tough.

But the meat producers are torn because we, the consumers, want our meat to be both cheap and delicious, and we can rarely have it both ways. For all the legions of meat scientists seeking out ways to keep us loving meat, their work is only partly about the flavors and textures. They are also working to find a perfect balance between the taste of meat and its price. Yet all too often meat that is cheap is of poor quality because it comes from animals that suffered much more than average, as if their fears leaked into the steaks and chops on our plates, making them tough and with an off flavor.

In a better world, all animals would be treated at least as well as Humphreys treats his "boys." In a better world, all meat would be juicy and tender and tasty—and still affordable. That's not the world we live in, or a world in which we will live in any near future. Although the taste of meat is important, for now it's price, not quality, that matters more in keeping us hooked on meat—no matter how much we'd like to believe otherwise. For the meat industry, keeping meat's price low is one of the time-tested ways to ensure we will continue buying its products. But, as we shall see, the meat industry has other ways, too: devising savvy marketing and advertising, lobbying, threatening lawsuits, sponsoring research, and intimidating critics.

6

Wagging the Dog of Demand

Sleek seems like a good word to describe the offices of the National Chicken Council (NCC) in Washington, DC. The sleekness begins on the street. The building at 1152 Fifteenth Street, which houses the NCC, is ultramodern, enclosed in glass, with a lavish lobby that echoes my footsteps as I walk in. Up on the fourth floor, I'm greeted by NCC's senior vice president Bill Roenigk—a jovial man who looks exactly like a "Bill." Roenigk leads me into a conference room where we sit at a large, oval table. Even though there are some vintage photos of farmers on the walls, the whole place feels much more "consulting" than "farming."

The NCC, just like its beef and pork equivalents (the National Cattlemen's Beef Association and the National Pork Board), is a trade association of meat producers. These organizations protect the interests of the industry, deal with PR crises, lobby the government, and arrange marketing campaigns. But at its core, their goal is rather simple: make sure Americans buy as much chicken, beef, and pork as possible. In other countries, similar organizations exist: the British Meat Processors Association, the Canadian Cattlemen's Association, and so on. Such organizations, together with powerful meat companies such as Tyson Foods or JBS, spend billions of dollars a year on lobbying and promotion so that we don't lose our appetites for animal protein. Some researchers argue that increasing meat consumption around the globe, the US included, is not demand driven but supply driven: it's pushed more by the actions of the meat industry and not so much by the desires of our taste buds. The industry doesn't exactly pretend otherwise, either. As the cattlemen's magazine *Beef* admitted in 2013: "The beef industry has worked hard to create the love affair that Americans have with a big, juicy ribeye."

The meat industry is capable of swinging our food preferences because it is ultrapowerful and ultraconsolidated. Consider these numbers: in 2011, in the US alone, the annual sales of meat were worth $186 billion. That's more than the GDP of Hungary or Ukraine. Moreover, just four pork producers control two-thirds of the market, and the top four in beef have about 75 percent of the market. Tyson, the largest meat corporation in the US, recently had a revenue of $34 billion—that's over *twenty times* as much as the GDP of Belize.

Other companies besides those that raise, slaughter, and sell meat benefit from consumers' carnivorous appetites: the fertilizer and pesticide producers, farm equipment manufacturers, seed growers (including Monsanto), soy and corn farmers, and pharmaceutical corporations, which sell antibiotics, beta-adrenergic agonists, and other drugs to the meat companies. In a way, they are all part of the meat business too. According to the American Meat Institute, the industry's primary trade organization: "Meat and poultry industry impacts firms in all 509 sectors of the U.S. economy. . . . The meat and poultry industry's economic ripple effect generates $864.2 billion annually to the U.S. economy, or roughly 6 percent of the entire GDP."

Compared to the meat industry, the vegetable and fruit industry has little clout. For one thing, if the name "vegetable and fruit industry" sounds odd, that's because such an expression is almost never used. The vegetable and fruit industry hardly exists as a united entity. In North America or the UK, only about five different types of meat really count in terms of sales: beef (including veal), pork, chicken, turkey, and lamb or mutton. Now think of all the different kinds of veggies, fruits, beans, and lentils out there. Or just consider the varieties of beans grown and sold in the US: pinto, navy, black, great northern, garbanzo, red kidney, lima, yellow eye, fava, mung, adzuki, marrow, appaloosa, anasazi.

The list goes on. Do lima bean producers want you to eat more lima beans? Of course they do. But they not only have to compete with garbanzo bean producers but also with other bean, pea, lentil, and vegetable growers. In a similar fashion, apples go up against peaches, blueberries against cherries. Even if the fruit and vegetable producers did unite, their sales would still be much smaller than those of the meat industry: in 2011, for example, all vegetables, fruits, and nuts combined made just over $45 billion in farm cash receipts. That's almost four times less than the livestock products earned. Beans, peas, and lentils—which are considered proper meat substitutes—fare even worse. In 2011 they made a staggering 140 times less than livestock products. Who has the power to convince you to love their foods and to eat more and more of them? Not the chickpea industry, that's for sure.

To make certain you keep eating meat, the industry levies almost a tax on products sold, known as beef and pork checkoffs. In the US each beef producer pays $1 per bovine head at the time the animal is sold, and each pork producer foots $0.40 per $100 of value. In Canada, the levy is $1 per animal head, and in Australia, it's $5 a head. Between 1987 and 2013, the US beef checkoff collected $1.2 billion, an impressive pile of money that is used "to increase domestic and/or international demand for beef"—in the words of the industry itself. To give you some perspective: one of the very few campaigns drafted to promote eating veggies, 5 A Day for Better Health, developed by the National Cancer Institute and the Produce for Better Health Foundation, had in 1999 a public communications budget of less than $3 million.

When Americans ask, "What's for dinner?," most will automatically reply: "Beef." That's hardly a surprise. Back in 1992 the industry spent $42 million of beef checkoff money spreading the slogan "Beef. It's What's for Dinner." As for its effectiveness, consider this quote from the industry's own website: "In the minds of the many consumers hearing that question ['What's for dinner?'], a dominant answer has been planted: Beef. It's what's for dinner. Not just planted, in fact. Watered, nourished and cared for over the past two decades." In 2015 the beef industry was planning to spend $39 million of checkoff revenues on promotion and research, "consumer public relations," "nutrition-influencer relations," and countering "misinformation from anti-beef groups." One of the checkoff websites, www.beefretail.org, is full of ideas on how to make people buy (and eat) more beef. Some examples: organizing cooking demonstrations on university campuses and student contests, providing in-store samplings of easy beef recipes, and employing influential chefs.

Yet it's the youngest consumers that the meat industry is particularly keen to hook on burgers and drumsticks. In their marketing effort, for example, they design "beef education" curriculums for K–12 classrooms. They are especially eager to attract Millennials, born between the early 1980s and early 2000s. To encourage them to eat more burgers and steaks, beef promoters share recipes on Facebook and use Twitter, Instagram, and Pinterest, where they post pictures of "delicious beef meals." They create apps and online resources: these, according to an industry marketing how-to guide, are "a must-have to attract and retain Millennials' interest."

These campaigns pay dividends. Between 2006 and 2013, every dollar dropped into the beef checkoff's piggybank returned over eleven dollars to the industry. If it wasn't for the checkoff, the industry has calculated that we would be eating 11.3 percent less beef. Meanwhile, the pork industry's "The Other White Meat" tagline is the fifth-most recognized ad slogan in the history of American advertising (the first being Allstate Insurance Company "You're in Good Hands"). During the five years following the campaign's kickoff in 1987, sales of pork shot up 20 percent. As C. W. Post, founder of General Foods, reportedly once said about cereal—but it could easily be said about meat: "You can't just manufacture cereal. You've got to get it halfway down the customer's throat through advertising. Then they've got to swallow it."

Checkoff programs are successful not only because they are large but also because the promotional messages of the meat industry are, according to the US Supreme Court, "government speech." These are not your typical marketing campaigns; they have the blessing of the government. The USDA actually reviews the promotional messages prepared by checkoffs. As David Robinson Simon writes in his book *Meatonomics*: "It may say it's the National Pork Board, but the background sounds you're hearing are the imposing bass tones of the US government . . . a lack of government involvement would likely lead to the decline—or maybe the end—of checkoffs."

Although the American poultry industry doesn't have a checkoff program, it still works hard to increase the demand for chicken and turkey. As Bill Roenigk explained to me, leaning back in his chair in the polished conference room of the NCC, meat demand is like a large dog, just sitting there, pretty immobile. But this dog also has a rather big tail. Good promotion and advertising is like grabbing this tail and wag-wag-wagging the dog as hard as you can. "So how do chicken producers wag the dog?" I ask. Roenigk laughs. "Social media campaigns are big at the moment," he tells me. "We are making September the 'Eat Chicken Month,' for example."

Yet generic promotion by meat producers—whether of beef, pork, or chicken—is just a part of the story of how publicity keeps us craving meat. Meat vendors, such as restaurants, also play a large part. Take McDonald's. Although it's not a meat company per se, McDonald's is the largest beef buyer not only in the US but in other countries, too (Ireland, for example). Selling on average about seventy-five burgers *per second* across the globe, McDonald's plays a large role in our ongoing love affair with meat. In 2011 it spent a whopping $1.37 billion on advertising. There are only thirty-six companies in the US that shell out upward of $1 billion a year on ads (think GM, Google, Apple). And no, veggie and fruit producers didn't make the list, unless you count Unilever with its soups and ketchup. And guess which ads appear most frequently on children's Saturday morning television? Number one: McDonald's. Number two: Burger King. The only figure that American kids recognize better than Ronald McDonald is Santa Claus.

Selling meat with advertising comes with a few simple rules of thumb. "Don't show animals" is a major one. A study done in Europe found out that it's better to avoid using any photos or even drawings of cows, pigs, or chickens, no matter how cute. "Rather to make the consumer reflect about the living animal, communication should be centered on other attributes linked to the hedonic sides of meal preparation and consumption," write the authors. And that is why you won't see many animals in meat ads. In other words: They don't want you to think about the animal too much or you may lose your appetite. But advertising, no matter how successful, is not the only way to ensure that demand for meat doesn't go down; there is lobbying, too.

Just one block away from the offices of the NCC is K Street: a stately line of heavy-looking buildings, among them steak houses and banks. There is nothing frilly on K Street, nothing cute, hipster, or kiddie friendly. It's a street of coffee-wielding suit-clad people, all in a rush. Consultants, lawyers, and, most of all, lobbyists—so many of the latter work here that K Street has been dubbed "the lobbyists' boulevard." Lobbying, as Roenigk tells me, is something that the NCC is now "focusing on." It's perfectly legal, of course, and involves arranging campaign contributions, encouraging lawsuits, and organizing public-relations campaigns—all to influence government policy. The Center for Responsive Politics estimates that during the 2013 election cycle, the animal products industry contributed $17.5 million to federal candidates. And such contributions appear to work. One study confirmed that changes in contributions do change voting behavior and that you can basically "'buy' legislators' votes" without breaking the law.

One thing that the meat industry would rather not lose (and would likely lobby intensely for if they were to) is government subsidies. According to Chuck Conner, deputy secretary of agriculture, producers of fresh fruits and vegetables "have traditionally been under-represented in farm bill policy." Meanwhile, between 1995 and 2012, American taxpayers helped pay $4.1 billion in livestock subsidies. It's a big number, but in reality what animal food producers actually receive—indirectly—is far more than that. The author of *Meatonomics* calculated that each year the US spends $38 billion to subsidize meat, fish, eggs, and dairy. Why is that number so much higher than the official livestock subsidies? One reason is feed grain subsidies. From 1995 to 2012, corn producers pocketed over $84 billion, and soybean growers $27 billion—which makes it considerably cheaper to buy corn and soybeans than to grow them. Since 60 percent of the corn and almost half of the soybeans that sprout from American soil are used for feeding livestock, subsidizing these crops is, to a large extent, tantamount to subsidizing the meat industry—and encouraging meat consumption.

If it wasn't for subsidies, we would be paying considerably more for our steaks and drumsticks. That would quite likely dampen our love affair with meat. "NCC did a study a couple of years ago," Roenigk tells me. "If you take the price of chicken and consumers' income, these two factors can explain 90 percent of why we eat more or less chicken." Imagine if beef prices were to go up 10 percent, would you buy less? Switch to chicken? Studies show that on average, a 10 percent increase in beef's price means about a 7.5 percent decrease in consumption. And then the demand for pork rises 3 percent and for chicken 2.4 percent—so it's bye-bye beef stew, hello chicken fajitas. However, some consumers, when faced with a higher bill at the butcher's, end up cutting down on meat altogether. In one NCC survey, 35 percent of shoppers said that when chicken prices increase, they just eat more veggies.

But the government is not the only entity subsidizing the meat industry. There are hidden costs to meat production that, instead of being paid by producers, are paid by taxpayers as part of what some call "subsidies by omission," and these costs are quite substantial. Neal Barnard, professor of medicine at George Washington University, calculated that, in 1992, direct health-care costs attributable to meat eating in the US were over $61 billion—from hypertension, heart disease, cancer, diabetes, and so on. In *Meatonomics*, Simon estimates that external costs of the animal food industry add up to at least $414 billion yearly—not only in health care but also in environmental costs such as pollution. For every dollar of beef or chicken sold, Simon argues, the industry imposes $1.70 of externalities on us. (In economics, an externality is a cost that affects a party that did not choose to incur that cost and which is not reflected in the cost of the goods.) So the next time you buy $10 worth of steak think about this: you are in reality paying $27 for it, just in installments—part at the checkout counter, part with your taxes, part with your health insurance.

What all such meat subsidies mean in practice is that for many Americans who struggle to make ends meet, buying a few burgers at McDonald's is often a cheaper way to feed the family than serving them lentil dal and a fresh salad. They may be eating meat simply because it's relatively low cost and readily available. As far as the meat industry goes, that's perfectly fine, of course. They want to keep the subsidies flowing and the externalities external. What they don't want is for the government to promote a plant-based diet.

On June 3, 2013, a seemingly trivial sign appeared at one of the food stations in the white expanse of the Longworth cafeteria—the bright, open space located in the same building as the offices of the House Agriculture Committee in Washington, DC. It's here on workdays around noon that a long line of staffers forms: members of Congress and lobbyists await their turn to grab lunch. On that particular Monday, many of them spotted a new sign advertising one of the food options. The sign, supposedly placed by one of the cafeteria's employees, simply said: "Meatless Monday." That was enough for the meat industry to raise an outcry. On June 7, the Farm Animal Welfare Coalition, a group that includes some of the nation's largest farm and ranch organizations, issued a statement to the House Administration Committee, protesting the appearance of the sign. In their letter, they wrote: "'Meatless Mondays' is an acknowledged tool of animal rights and environmental organizations who seek to publicly denigrate U.S. livestock and poultry production." On the following Monday, June 10, the "Meatless Monday" sign at the Longworth cafeteria was gone. It has never appeared again.

There is much more to the meat industry's pressure on the government than a disappearing "Meatless Monday" sign, of course. As Marion Nestle, professor of nutrition at New York University, has stressed, the meat industry has, in recent years, won the major battles. One of those battles has been over the Dietary Guidelines. The guidelines, according to the USDA and the US Department of Health and Human Services, which jointly publish them every five years, "provide authoritative advice . . . about consuming fewer calories, making informed food choices . . . to promote overall health." Nestle has a different definition of the Dietary Guidelines, though. In her book *Food Politics*, she writes: "Dietary guidelines are political compromises between what science tells us about nutrition and what is good for the food industry." Among the words and phrases that the meat industry doesn't like are "eat less," as in, "eat less meat." Over the years the standard word used by the

Dietary Guidelines has been *choose* ("choose lean meat") instead of "eat less." *Choose* doesn't bother the industry as much because it encourages people to go out and buy more chicken or less fatty beef. Another standard tactic is to point a finger at particular nutrients but not the foods that contain them. So it's "no" to cholesterol and fat but silence about fatty meats. On the first day of her job working on the editorial production of the Surgeon General's Report on Nutrition and Health, back in the 1980s, Nestle was given clear rules. She recalls: "No matter what the research indicated, the report could not recommend 'eat less meat' as a way to reduce intake of saturated fat. . . . When released in 1988, the Surgeon General's Report recommended 'choose lean meats.'"

How is the meat industry able to put so much pressure on the Dietary Guidelines committees and the USDA? The first issue lies with the function of the USDA itself—and goes back to the establishment of the department in 1862. Since its very beginnings, the USDA has had two roles: one was to help the industry achieve a reliable food supply and sell more products and the second was to advise Americans on their diets. And here is the problem. Today—as opposed to the nineteenth century when undernutrition was the challenge—those two roles just don't fit together well. The USDA has a conflict of interest at its very core.

The second issue also concerns conflicts of interest, but this time among members of the committees drafting dietary recommendations. Over the years, some guidelines committee members received grant supports from the National Live Stock and Meat Board, served on the grant review committee for the American Meat Institute, or had their research supported by the National Dairy Council—to name but a few. And then there is the phenomenon of revolving doors: industry people changing careers to become government people, and vice versa. Examples? Dale Moore, chief of staff to Ann Veneman, the secretary of agriculture, was formerly the executive director for legislative affairs of the National Cattlemen's Beef Association (NCBA). Deputy Secretary James Moseley co-owned a large factory farm in Indiana. USDA director of communications Alisa Harrison used to be NCBA's executive director of public relations, while NCBA's former president, JoAnn Smith, got appointed as chief of USDA's Food Marketing and Inspection Division. The list goes on.

There is one more place where meat-industry-related conflicts of interest pop up: scientific research. If you scroll down to the author disclosure sections of research papers published in peer-reviewed journals, in some of them you may discover that the scientists behind the study received funding from the meat industry. For example, the author of one 2012 analysis, which praises beef as a great source of protein, "has been paid by the Beef Checkoff . . . to provide consulting services." The author of yet another research paper, published in 2014, and claiming that lean beef has benefits for cardiovascular health, "received grant funds from the Beef Checkoff Program." Sometimes the connections may be pretty obscure. A Swedish study, which is often quoted as proof that vegetarian diets are unhealthy, was supported by a grant from the benign-sounding Swedish Nutrition Foundation. But if you go to the foundation's website, you may find out that there are several meat and dairy businesses among its "member companies," including McDonald's. Besides sponsoring scientists directly, the meat industry sponsors organizations promoting good nutrition. Tyson Foods, the California Beef Council, and the Texas Beef Council, among others, give money to the American Heart Association. The American Dietetic Association Foundation receives funds from the National Cattlemen's Beef Association, and so does the Academy of Nutrition and Dietetics Foundation.

Of course, the fact that someone receives funding from the meat industry doesn't automatically mean that his or her research will be skewed to cheer the consumption of meat. But it may. In 2013 editors of several scientific journals, including the prestigious *BMJ* (formerly the *British Medical Journal*), announced that they will no longer accept studies funded by the tobacco industry. The editors wrote that although some may say that funding doesn't equal endorsement, such a view "ignores the growing body of evidence that biases and research misconduct are often impossible to detect." Research done on the pharmaceutical industry, for example, found that sponsored studies were up to four times more likely to be favorable to the sponsor than studies that didn't receive such funding. And favorable research is important to ensure that demand for meat is strong. According to the pork industry itself, a 10 percent increase in "demand enhancing research" increases per capita pork demand by 0.06 percent, controlling for everything else. This may seem like a small number, but with pork sales of $97 billion, 0.06 percent is a respectable $58 million a year. In other words: it's a game worth playing.

On the other hand, some scientists who study vegetarian nutrition and publish results that suggest we should "eat less" meat suffer under pressure from the livestock industry. That's a story I've heard over and over in my interviews, including from a top United Nations nutritionist and from T. Colin Campbell, professor emeritus of nutritional biochemistry at Cornell University. Campbell began his career assuming that he would research the benefits of meat consumption (he hunted, fished, and devoured tons of meat himself), but he soon discovered that the data just wouldn't support the idea that meat-heavy diets are good for us. After decades of research, including the renowned China Project (called by the *New York Times* "the Grand Prix of epidemiology"), Campbell became one of the best-known advocates of plant-based eating. Yet he also suffered his share of adversities from the livestock industry. As he told me over the phone in a soft voice carrying a suggestion of his age (he is over eighty), not only did his vegetarian nutrition course at Cornell get cancelled by a senior administrator who was "personally deeply beholden to the dairy industry," even though the course was hugely popular with the students, but also, another time, the egg industry "contacted the dean of our college to have me dismissed." He didn't get fired, nor did he change his line of research—the incident actually encouraged him to do his work. But if Campbell had been a journalist and one of his stories had suggested that meat is not safe for human consumption, he might have ended in much more trouble. Just look at what happened to Oprah Winfrey.

On her show that aired on April 16, 1996, Winfrey had among her guests William Hueston from the USDA, Gary Weber from the National Cattlemen's Beef Association, and Howard Lyman, ex-rancher and meat lobbyist gone vegetarian, known currently as the "Mad Cowboy" (after the title of his book). They talked about a new variant of

Creutzfeldt-Jakob disease, which is fatal and causes sponge-like lesions in the human brain, and discussed its possible connection to eating beef infected with bovine spongiform encephalopathy. At some point in the show, on hearing about industry practices (as reported by Lyman), Winfrey exclaimed: "It has just stopped me cold from eating another burger." As a result, both Winfrey and Lyman got sued by a group of cattlemen under the Texas food libel law—for making customers afraid to eat beef.

"Burgers v. Oprah" (as the *New York Times* dubbed it) was the most famous case stemming from US agricultural disparagement laws. In thirteen states (Alabama, Arizona, Colorado, Florida, Georgia, Idaho, Louisiana, Mississippi, North Dakota, Ohio, Oklahoma, South Dakota, and Texas), suggesting that a food is not safe for human consumption can get you sued. If that suggestion is proven false, you lose. Any perishable food producer could take you to court for making statements about the hazards of its goods. But it appears that the livestock industry is more inclined to use food libel laws than, say, the cucumber or pineapple industry.

Although Winfrey and Lyman ended up winning the case, Winfrey's legal fees supposedly exceeded $1 million. Costs on this scale can successfully silence journalists and writers. As Robert Hatherill, author of *Eat to Beat Cancer*, stated in his op-ed piece for the *Los Angeles Times* in 1999: "My publisher stripped lengthy passages from my new book. Simply put, I was not allowed to disclose dangers inherent in some common foods like dairy and meat products. . . . The problem had nothing to do with whether there was sufficient evidence to support the claims—there is—it came down to fear of litigation. I was told, 'We could win the lawsuit, but it would cost us millions, and it's just not worth it.'"

When I talk to Lyman on the phone, once we get to the topic of "Burgers v. Oprah" (or rather, from his perspective, "Burgers v. Lyman"), his voice becomes stronger, his words flowing faster. "In a way the meat industry has won this lawsuit," he tells me. "Since then media is really afraid to say anything bad about the meat industry; they fear they will get sued." "Should I fear, too?" I ask him, only half jokingly. Lyman laughs. "Just get yourself really, really good insurance."

Food disparagement laws are not the only ones that may quiet potential critics of the meat industry. The other ones are the "ag-gag laws"—so christened by *New York Times* columnist Mark Bittman. Say you are an animal rights activist, or an undercover journalist, and want to find out the truth about practices on industrial farms. You apply for a job at one of those places and, once hired, come to work equipped with a hidden camera. Over the duration of your short employment, you may record things that are heart wrenching, brutal, and inhumane. In the past, investigations like that have produced videos showing workers bashing cows' heads with pickaxes, injured piglets with their legs duct-taped to their bodies, and fully conscious chickens boiled to death in feather-removal tanks. Yet exposing such barbarities is often illegal. In Iowa, for example, law HF 589 makes it practically impossible for journalists to get jobs at factory farms to document what's happening inside. According to this law, it's a criminal offense to give false statements on your farm or slaughterhouse job application. So farms and slaughterhouses ask simple questions on their application forms: Do you work for media or an animal rights organization? Are you in any way affiliated with them? If you answer yes, you won't get hired. If you lie, you end up in court—and later, possibly even in jail. Do ag-gag laws affect our appetites for meat? Try watching one of the undercover videos on YouTube (type in "undercover industrial farming") and find out for yourself whether your willingness to eat chicken or pork will change.

The policies of governments influence how much meat we put on our plates. They influence it through subsidies, dietary guidelines, and laws. At the same time, the meat industry advertises and promotes. It lobbies. It funds research. It sues. But it's hard to blame the industry for wanting to sell as much of its product as possible. It's their job, after all. As the meat magnate Phil Armour stated back in the nineteenth century: "We are here to make money. I wish I could make more." They are not here to care whether eating too much meat is bad for your arteries, whether it is bad for the planet, for the animals, or for feeding the hungry. But we consumers often seem to forget that meat "harvesting" is business, just like selling perfumes or shoes is. We forget that, to some degree, we love eating meat because it is very well *sold* to us. What the meat industry does is just wag the dog of demand as hard as it can.

Yet the dog wagging wouldn't go so smoothly if consumption of meat wasn't deeply entrenched in our culture and if the animal protein business couldn't play on the mighty symbolism that meat carries. As we shall see, meat keeps us hooked because for generations we've been linking it with power, wealth, and sex.

7

EATING SYMBOLS

The first time I smelled a durian fruit, I thought I caught a whiff of someone's dirty socks, or maybe of a dead rat. Yet in Singapore—a steam room of an island where I once lived—durians are loved by many. People there actually call it the "king of fruit." You can see the large, heavy fruits with their spiky husks piled up high on the street stalls of neighborhoods such as Geylang, among pawnshops, sex shops, and tiny restaurants where pigeons walk across tabletops and orange-beaked birds rummage through rotting garbage. I've never been bold enough to actually taste a durian, not since reading in one newspaper that its flavor brings to mind either vanilla or the taste of stale baby vomit. Still, all over Singapore, durians reign. Fancy bakeries sell durian cakes, upscale restaurants lure customers with durian ice cream, and even McDonald's introduced a durian McFlurry. What makes it possible for someone to enjoy something that to others can smell of dirty socks and taste of baby vomit? The answer is simple: culture.

What we eat and the way we eat it are only partially inscribed in our genes. We may have some preference for protein foods in our DNA, or plentiful taste buds that make us cringe when bitter veggies end up on our tongues, but that's only part of the story—and a rather small one at that. If you've ever been to a food store in a distant nation, you might have found things there that you wouldn't happily toss on your plate: donkey penises (China), ultrastinky fermented herring (Sweden), tuna eyeballs (Japan). Do people there eat these things because they have different taste buds? Because a powerful tuna eyeball industry tells them to do so through TV advertising? Not really.

If you took two people and wanted to guess what they liked to dine on, running tests on their taste buds wouldn't be the best way to go; the easier method would be to just ask for their passports. In one large study, children from several European countries were offered apple juice of varying sweetness and crackers that were either fatty or lean and with or without added salt. Nothing predicted the kids' taste preferences better than their place of birth—not the education of their parents, not how much TV they watched. Food, it appears, is a matter of culture. When we eat we swallow not only nutrients but also meanings and symbols. It's our society that teaches us what's edible and which foods should be desired. And in Singapore, one highly desirable food happens to be durian.

The lesson of what foods are good to eat begins in the mother's womb. The flavors from a pregnant woman's meals seep into her amniotic fluid, which the fetus swallows and tastes. Studies show, for instance, that if she eats a lot of carrots, her baby, once born, will enjoy carrot-flavored cereal more than kids whose moms didn't go carrot crazy during pregnancy. Later, this culinary education continues on the breast. The process is similar: what the mother eats flavors her milk, and that's what, in general, her child learns to prefer. If you don't care for aniseed much, for example, it's quite likely your mom didn't eat it in your earliest days.

Then comes weaning time. Human babies, along with other baby omnivores, like rat pups, acquire tastes for the same foods the adults in their lives enjoy most. As scientists put it, their food preferences are "socially transmitted." Since baby rats are not spoon-fed in high chairs, they learn what's good to eat by sniffing the mouths or feces of their older companions, where the scent of food may still linger. As a result, rats develop ethnic-like "cuisines," since two groups living in similar environments may have learned generation after generation to choose some foods over others. If a rat traveled from his home (say New York's Grand Central subway station) to another rat enclave (Times Square station, perhaps), he likely would be quite surprised by what the others lived on. But rats and humans are not the only animals whose culinary likes and dislikes vary with culture: it's the same with baboons, sparrows, lizards, and even fish. The foods that an animal learns to like from its parents can be quite bizarre, too. In experiments, kittens who saw their mothers munch on bananas acquired a taste for this rather unkitty food. If a child observes people around him eating hamburgers, fries, and ketchup, he will likely grow up to enjoy these particular foods, too (and he is probably American). If she grows up among lovers of mopane caterpillars, millet mush, and hippo meat, that is what she'll prefer as an adult (and she is probably Zambian).

When it comes to liking a food, whether fried bacon or durian, even the facial expressions of the people around us are important. Studies show that if someone makes a disgusted face when a child is eating, the young one may lose his appetite. Meanwhile, a happy smile can open tightly sealed toddler lips to things that were previously considered inedible. Of course, parents often subconsciously react to foods they serve their kids. If you hate brussels sprouts, it's hard not to wince while offering them to a child. If you love bacon, you may lick your lips with pleasure when the tiny fork cruises to your kid's mouth. Children as well as adults, being the social creatures that we are, learn what's yummy in social settings. We simply observe what others like and dislike and follow their lead. In Mexico, kids discover from a young age that the pain associated with chili peppers is considered a good thing, and so they start to enjoy it. We also learn to like best those foods that we eat during particularly pleasant occasions. Brian Wansink, an expert on eating habits from Cornell University, once described an example of how Asian exchange students who come to the US start seeing cookies as a comfort food. It goes like this: a Chinese, let's say a girl, like most others in her culture, is not used to eating cookies. She moves to the US for college. She goes to parties and sees people eating cookies and having fun. She goes to more parties and sees more people eating cookies and having fun. Time passes. Parties, fun, cookies, parties, fun, cookies. Soon the girl makes a subconscious connection between fun, social life, and cookies. And then one day when she feels lonely and sad, she buys herself a pack of Oreos, trying to re-create that feeling she had at the parties. It works, and she gets hooked on cookies.

Social sharing during feasts is one of the most important reasons why meat keeps such a tight grip on us. This impulse likely dates back to when our hominin ancestors were divvying up their kills. Meat was the perfect food to share for a celebration: it came in a big package, too big for a single person or even a family to down in one go, and it spoiled if not eaten fast. It was hard to get—and still is—either because it's difficult to hunt on a savanna or it's difficult to hunt down enough money in your wallet to buy it. For this reason, meat is *the* food for feasting, so much so that in many cultures people call celebrations a "time to eat meat." Even the word *carnival* comes from the Italian *carnevale*, meaning "good-bye meat" (as in, good-bye, see you after the fasting of Lent). Sharing food, and meat in particular, makes people feel they belong, whether it's grilling beef burgers with neighbors (North America), roasting pork kiełbasa over a campfire with friends (Poland), or drinking a bowl of *boshintang*, dog soup, with family during the hottest days of summer (South Korea). That's because we share not only the food but also the memories of fun. Like the cookies of the Chinese student, meat becomes even more valued because it gets connected to pleasure, to feelings of togetherness. No wonder then that when a vegetarian says "No thank you" to a cut of steak, dark clouds of social disapproval quickly gather. His hosts begin to think, "What do you mean, no thank you? No thank you for sharing food with us? You don't want to be a part of our group?"

Although we learn to eat and like meat because that's what the people around us do and have been doing for generations, because that is what our pregnant and breastfeeding mothers filled up on, and because that's what they spoon-fed us in our toddlerhoods, meat is much more than a cultural habit. Its allure stems also from the powerful symbolism that it carries: it embodies strength, masculinity, wealth, and dominance. We eat meat in part because of the unconscious assumption that we are what we eat.

I'm standing on the packed, brownish dirt that is the floor of the Temple of Pythons in Ouidah, Benin. Paul Akakpo, my guide to West African voodoo, adjusts the large python that is wrapped, jewelry-like, around his neck. Akakpo has a remarkable knowledge of everything voodoo: not only is he a practitioner himself, he is also the nephew of a head priest of Beninese voodoo (a voodoo "pope," Akakpo tells me). As we talk, Akakpo points to the ground, to the temple's "floor." I follow his gaze and shudder: there is blood under my feet.

A trail of red winds around the concrete steps of the temple and ends by a giant sacred tree, where a local woman in a Technicolor dress is holding a freshly killed chicken, smearing its warm blood on the trunk of the tree. She is feeding the spirits. The woman's shoulders are naked, her dress starting barely at the top of her breasts. To all in the know, those naked shoulders signify she is a voodoo priestess. When she is done with the chicken, she throws its limp body on the ground and wipes her hands on a piece of cloth. She doesn't drink the blood, not now, not when I look at her. Maybe she will later? "Blood and meat eating is a direct communion with the voodoo spirits," Akakpo explains. "It empowers the believers with the voodoo spirits and helps them conquer their enemies." Voodoo adepts drink the blood and eat the meat of sacrificial animals because they believe it gives them the power of the animal, its strength.

There is a reason why people believe mastering other creatures gives us their powers. Animals are dangerous and difficult to kill. Something as small as a rabbit can scratch you and inflict wounds that may putrefy. On the other hand: Do you know anyone attacked by a cabbage? Exactly. That's why in our collective human mind (and not just the West African one) animals and blood mean strength and aggression. And since we are what we eat, as the saying goes, eating animals makes us powerful, resilient, tough. It's not just the voodoo believers of Benin who take this literally—sometimes Americans do as well.

That's at least what Paul Rozin, professor of cultural psychology at the University of Pennsylvania and a world-class expert on human food choice, discovered in his experiments. Rozin, who in the 1970s coined the term *omnivore's dilemma*, believes that culture is a powerful determinant of our culinary choices. In one of his experiments, he let a group of students read about an imaginary distant society in which people frequently dine on boars, while a second group got a similar text about a culture of turtle eaters. Afterward, when Rozin asked the students to describe a typical member of each society, the boar eaters were characterized as more aggressive than turtle eaters, faster, and more hairy. That's a perfect example of "you are what you eat": if you consume boars, you become boar-like.

What makes this belief even more powerful is that sometimes it really *is* true. If you eat a lot of carrots, the carotene in them can color your skin orange. If you eat a lot of fat, you become fat. From there it's easy to imagine that consuming horses or bulls could make you powerful and eating lettuce could make you slack. Our language reinforces that connection. To "beef up" is to make something stronger. A "vegetable" is a severely impaired person. We have a "couch potato," not a "couch steak." We "veg out." And since few of us want to be slow and weak, we would rather consume animals than vegetables. That is particularly true of men.

You may have seen this commercial while you were "vegging out" in front of the TV: a young man in a green polo shirt is waiting at the grocery checkout, when another guy, more or less his age, stands in line behind him and starts unloading his shopping cart onto the conveyor belt. The cashier scans the first guy's groceries: tons of green stuff, some radishes, and tofu. While she beeps the stuff away, the tofu man glances at the guy behind him and takes in his shopping: red, chunky ribs, piles of some other unidentified meat. Meat, meat, and more meat. The tofu man begins to look uncomfortable, as if his shoes are too small or his collar is too tight. Then his eyes land on an advertisement for a Hummer H3. He pushes his veggie-laden cart out of the store, his face purposeful now, determined. He drives straight to a GM dealership and, without hesitation, buys himself a Hummer. As powerful music rocks in the background, the tofu guy drives away in his new freight container of a car, nibbling on a carrot. Big bright letters pop on the screen: "Restore the balance."

What I've just described is a Hummer H3 spot that aired in the US in 2006. The original tag was actually "Restore your manhood"—but people complained, so it got changed. Yet the message remained pretty clear: real men eat meat, and if you don't, you can at least boost your masculinity by driving a huge, earth-unfriendly car.

You don't have to look far to find ads claiming that men need their meat. Domino's Pizza did one with a similar message and so did Taco

Bell, McDonald's, Jack in the Box, Quiznos, and TGI Fridays. In Burger King's commercial "Manthem," guys are "too hungry to settle for chick food" and need a Texas Double Whopper to "eat like a man." In New Zealand, a campaign promoting Lion Red beer gives guys "Man Points" for doing masculine things (building a deck, fishing with pals, starting a barbecue) and negative "Man Points" for asking for directions, waxing anything that's not a board, or cooking tofu.

Teachers of introductory sociology classes sometimes recommend their students do an experiment. Head to a restaurant on a date with the opposite sex. The guy should order veggies; the girl should ask for a steak. Now observe the waiter. In all likelihood, he will get the food all mixed up and place the meat in front of the guy. But it's not just waiters who think of meat as masculine—college students do that, too. In another of Rozin's studies, when asked which foods are the most "male," University of Pennsylvania students chose steaks, hamburgers, and beef chili. The top "female food," meanwhile, was chocolate.

The notion that meat is *the* food for men is nothing new, of course. One painting of Henry VIII portrayed him eating steak, while his six wives nibbled on apples, turnips, and carrots. This dichotomy—meat/male, veggies/female—was particularly pronounced during wars. If meat could make men stronger, no one needed it more than soldiers. A Tudor knight received two pounds of meat a day in provisions, while scores of his countrymen almost never ate any animal flesh at all. Even much later, during World War II, American GIs consumed 2.5 times as much meat as civilians back home. Animal protein, those in command believed, was necessary for superior frontline performance. Like the West African voodoo practitioners, the American war planners thought that meat eating could help conquer the enemy. If you were to fight like a lion, you had to eat like a lion, too.

To find out the roots of this belief, it's worth asking who it was that most likely started the urban legend (or rather savanna legend) that eating meat makes you assimilate the powers of the animal. The answer, most likely, is that men did. They were the ones who brought home the rare treat of a mammoth sirloin or a giraffe filet mignon. They were the ones who decided how to divide the meat and who could receive a share.

They gathered around campfires (as soon as they were invented, that is), just as they gather around $2,000 stainless-steel gas grills today, and talked politics. Hunting and eating meat reinforced gender inequality. To make sure that women didn't get the powers of animals from their meat or challenge the position of men as providers of that rare but nutritious food, taboos were put in place. Even today, most meat taboos are directed at women. Some African tribes, for example, forbid them to consume chicken, while others, like Tanzania's Hadza, reserve the fattiest portions of game for men. If a woman dares steal a bite, she risks rape or even death. Such taboos, of course, helped men secure the stomach-filling bits for themselves.

Generation after generation, meat's tie to masculine identity was reinforced, becoming an expression of a patriarchal world. What's more, sex got added to the mix. Today some people may jokingly celebrate "National Steak and Blow Job Day" on March 14 (and yes, it's real), but in the past, the connection between the consumption of animal flesh and lovemaking was taken far more seriously. In Victorian times, meat was believed to drive lust, and schoolboys were advised to give up eating it so that they would stop masturbating. Animal flesh was thought to be "too strong" for pregnant women and to lead young girls to nymphomania. The irony is, recent scientific data demonstrates that the opposite may be true and that meat may not be the sexual tonic that our ancestors have claimed. Research shows that frequent intake of meat may negatively affect semen quality (which also casts a shadow on the premise that "meat is for real men"). What's more, if a guy's mother frequently ate beef when she was pregnant with him, as an adult he may have a lower sperm concentration than the sons of less beef-loving women.

Why would meat and sex go together at all, though? According to Carol J. Adams, our patriarchal society has forged that connection. A feminist, a writer, and, as she calls herself, "an activist immersed in theory," Adams became famous after the publication of her book *The Sexual Politics of Meat* back in 1990. She stepped on many toes with that book. The British *Sunday Telegraph* joked that it was actually written by "a male academic emigré from Eastern Europe, who poses as a madwoman." Adams is not an academic, nor a male Eastern European. She is also far from mad. When I call her for an interview, she recommends I turn on a recorder: "I speak fast," she warns me. And it's true. She has a lot to tell me, her thoughts flowing one into another.

Adams believes that in a patriarchal world both animals and women are treated not as subjects but as objects. For animals, that often means they end up eaten; for women, that means they are second-class citizens who also sometimes get sexually "consumed" against their wills. Adams explains: "Meat eating benefits from objectification in a way similar to sexual violence because you don't see the other being as a living, breathing individual." Through the distance that objectification provides, men are taught to look down on women, and everyone is taught to stay hooked on meat. One survey of one hundred nontechnological cultures found that the more a tribe bases its diet on animal products, the less power women hold. Also, in an interesting twist, the more meat is consumed in a society, the more distance fathers keep from their infants. The stereotypical twenty-first century BabyBjörn-wearing father (who also likely took leave from work to stay home with the baby) may be a vegetarian, too.

Yet don't let the apparent ubiquity of BabyBjörn-wearing, tofu-loving fathers fool you. As Adams tells me, the connection among sex, meat, and men has only gotten stronger since she has published her book. "Back in the '80s we had some success with feminism, the animal rights movement was beginning to strengthen, and I felt that maybe I'll finish my book just in time to be commenting on something that was passing away," she says. But it didn't pass away. Today, TV ads persuade men that they need meat to be masculine—and so does the popular press. The lifestyle magazine *Men's Health*, which has a US circulation of 1.65 million, is particularly vocal about this. A *Men's Health* ideal guy eats tons of red meat. "Vegetables are for girls . . . If your instincts tell you following a vegetarian diet isn't manly, you're right," stated one article.

So where does this outpouring of meat-eating machismo come from? Adams tells me that traditional masculinity is threatened nowadays by feminism, the gay movement, metrosexuality, and all the BabyBjörn-wearing carrot-munching fathers of the world. Old-school masculinity needs to be reaffirmed, and one way to do this is to connect it once again with eating bloody slabs of animal flesh, even if that flesh didn't require any skills or strength to kill and came in a plastic wrap from a supermarket. Twenty-first-century men may feel they are losing their power and dominance, and they want it back. Adams is not alone in this belief. Other researchers, too, point to this "crisis of masculinity" and see eating meat as a symbol of returning to manhood's roots. On the other hand, to reject meat is to reject the mainstream notions of masculinity and, in a way, patriarchal society itself. Men who do this risk ridicule and opposition from the other, steak-loving males. As Adams wrote in her book: "They are opting for women's food. How dare they?" Meanwhile, for women, giving up meat can be a way to separate themselves from a traditional, male-dominated society in which both women and animals are objectified. That is likely why many nineteenth-century suffragists were vegetarian, and why today Adams advocates that feminism and vegetarianism should go hand in hand and help each other out. On the flip side, for some women, eating bloody roasts may be a stand-in for joining the ranks of the powerful (think dining on a medium-rare steak in one of Washington, DC's ubiquitous steak houses). If they can't have the whole world, they at least want a bite. That makes perfect sense also because meat symbolizes power over the poor, underprivileged masses. From our earliest days on the Paleolithic savanna, when our ancestors were showing off their kills to form alliances and gain social position, meat has always stood for luxury and for riches.

It was November 1922 when an archaeologist named Howard Carter discovered the tomb of Pharaoh Tutankhamen, which had lain undisturbed in the scorchingly hot Egyptian sands for over three thousand years. Once Carter made a small hole in the blocked doorway of the tomb's antechamber, he squinted inside. Among the first things he spotted were forty-eight whitewashed, wooden cases. As he soon learned, they were full of meat. But the various joints of beef and poultry weren't just dumped into the boxes straight from the butcher's: if they had been, they would have spoiled into a stinky mess pretty fast. Instead, they were carefully treated with balms, in a manner similar to that of mummies. As of today, archaeologists have found hundreds of such ancient "meat mummies"—carefully preserved meat that high-status Egyptians took with them to the afterlife. Already back then, it appears, meat was a symbol of wealth, both in this life and in the netherworld.

For centuries, animal flesh was a perfect indicator of how rich someone was. To be a symbol of wealth, an object has to be rare and difficult to obtain. Think Patek Philippe watches versus T-shirts from Walmart. Maybe you don't have to risk your life chasing down a fancy watch on a savanna, but if you are an average American, you would have to spend many hours at work to earn the thousands of dollars necessary to pay for it. Wealthy people use Patek Philippe watches and other expensive things to set themselves apart from the masses. Such items need to be pricey and hard to obtain. One joke I've recently heard sums it up pretty well: A Russian oligarch meets another Russian oligarch at a party and admires his tie. "Silk?" he asks. "Silk," nods the other. "How much did you pay?" "$5,000," says the tie's owner. "Really? That's a bad deal," the other guy shakes his head. "I've seen exactly the same tie in a boutique in Moscow. You could have easily gotten it for $10,000!"

In the past, meat used to be like those silk ties—and still is in many parts of the world. It was desired because it was hard to get and expensive. Psychological experiments show that if shop owners advertise that something is available "only today!" or that there are "only ten left!," people will want that thing more, even if they don't really need it. In medieval Europe, turnips, cabbage, and beetroots were common—so there wasn't much reason to crave them. Meat, on the other hand, was such a rare treat that peasants barely ever ate it. Meanwhile, the aristocracy could down as much as three pounds per person *per day*. When Henry IV, the king of England, married Joan of Navarre in 1403, their wedding feast was loaded with meats. There was a boar's head, pheasants, heron, "calves foot jelly with white wine and vinegar," stuffed suckling pigs, "peacocks served in their plumage," cranes, quails, young rabbits, and rissoles of pork roasted on a spit. The kings and queens of Europe, just like the rest of the aristocracy, ate hardly any fruits and vegetables. One British "shopping" list for an elite dinner party of fifty included thirty-six chickens, nine rabbits, four geese, one swan, two beef rumps, six quails, bacon and fifty eggs, some spices, and little else. As for spices, people in medieval Europe didn't use them to cover the taste of animal flesh gone bad. That's a myth. Back then if you could afford spices, that meant you could afford fresh meat and could throw out whatever got spoiled. Spices were just another indicator of wealth that the masses couldn't dream of affording.

Some animals were a status symbol even before they became meat. Take cattle. Owning a cow that gave you milk or an ox that ploughed the fields (which could be eaten once they got too old to work) was extremely valuable, so much so that in several European languages the word *cattle* is synonymous with "capital." In Sanskrit, the word for battle (*gavisti*) basically means "desire for cattle." The more cows and oxen you could afford to slaughter—to waste, in a way—the more powerful it meant you were. Killing a cow for a feast to share with others showed you were rich, that you'd "made it." Even today in many African or Latin American cultures, wealth is measured by how many head of cattle one owns. I experienced this personally when in my early twenties I traveled to Tanzania with my stepfather. One day when we were strolling through a market, a local man approached us. He took my stepfather to the side and in all seriousness offered him four cows in exchange for my hand in marriage. Four cows was a generous offer, he said, which meant he was wealthy enough to take a European wife. The offer wasn't accepted (of course), but the equation—wealthy Tanzanian equals lots of cows—stuck in my mind.

Although modern Westerners are rather unlikely to show off their affluence by the amount of cattle they own (unless they are farmers or meat industry magnates) or by the number of boars' heads they serve at a party, they may try to impress others by the price tags of their grills. We may like grilling because of the way it reminds us of cozy Paleolithic campfires or because the Maillard reaction boosts the flavors of broiled meats, but that's just part of the story. Our penchant for roasting and the way we value it over boiling once again goes back to the symbolism of meat as the food for the rich and powerful. The thing is, you can't really roast a low-quality cut from an old milked-out cow. It won't be any good—just chewy and tough. To make such meat edible, you'd have to stew it for a long time. Boiling not only makes it possible to dine on inferior cuts, but it also preserves juices, making this particular cooking method more economical. It's a good way to prepare salted meats, too, or ones that are not exactly fresh. Ergo, boiling is perfect for the poor and the hungry. For roasting you need young animals and premium cuts. This gave the aristocracy another opportunity for displaying their wealth: look, we can afford to roast all these freshly killed, tender younglings. The poor could spare meat to roast only on special occasions. Thus grilling became associated with wealth and with celebrations, and that is also why grilling is for guys and stooping over a pot of stew is for women—the first one is prestigious, the second one is not.

The association with wealth is also one of the reasons why Americans are so into beef and not so much into pork. First, many settlers came from Great Britain, where beef was a well-established food for the mighty and powerful. To be just like the nobility back home, the new Americans wanted steaks. Second, pork was considered a meat for the poor. Hogs were cheap to raise: they could basically feed themselves by eating garbage off the streets, which they commonly did. Packs of swine roamed the American cities, including Boston, Philadelphia, and New York, well into the nineteenth century. Pork was easier to preserve than beef, too, which meant that before the advent of refrigerators, the lower classes often relied on salted, barreled pig meat to survive winters. Up until the early twentieth century, it was pork that fed America. But it was expensive, rare, and tough-to-cook beef that people craved—precisely because it was expensive, rare, and tough to cook. That's human psychology 101.

Just as it has long symbolized power over women and over the poorer members of society, meat has also long stood for power over other, less affluent nations. Food is a potent marker of ethnic identity. When immigrants move to a new country, they may start speaking the new language and pack away their traditional clothes in the attic to gather dust, but the foods of home are among the last things to go. One of my Singaporean friends living in France has so much Asian chow stashed under her bed she could survive World War III living on it (thankfully, she doesn't have any durians in her stock). Food is commonly used for setting nations apart: we call the Germans "krauts" and the French "frogs." The idea that ethnic cuisine heavy in meat makes for better citizens became popular in the nineteenth century. George Miller Beard, a physician quite well-known back then in America, wrote eloquently on the subject: "Savages who feed on poor food are poor savages, and intellectually far inferior to the beef-eaters of any race. . . . The rice-eating Hindoo and Chinese and the potato-eating Irish peasant are kept in subjection by the well-fed English." This belief that a meat-eating nation equals a better nation held strong in the West until the mid-twentieth century. In a book published in 1939 by Swift & Company (a meatpacking corporation), that point was made quite clear: "We know meat-eating races have been and are leaders in the progress made by mankind in its upward struggle through the ages." Even after the war, a textbook for butchery students hailed that "the virile Australian race is a typical example of heavy meat-eaters." Once again, to eat meat is to be powerful. This is a variation on the "you are what you eat" theme and on the "we are better than you because we can afford something as expensive as meat" theme. Meanwhile, by rejecting meat a vegetarian rejects not only his "tribe" at the table but also often, in a way, his whole nation. The equation of a Briton as an eater of beef was already in place in 1542, when Andrew Boorde recommended beef as the perfect food for Englishmen in his guidebook *Dyetary*. On one British ship, the *Titanic*, dinner was announced by the sounds of "The Roast Beef of Old England," and the personification of UK is the beef-loving John Bull. Now imagine you are English and you stop eating beef. That's a lot of heritage to reject, a lot of national identity to part with. If everyone in Great Britain switched to plant-based diets, should the country's personification be renamed John Veggie?

In the US, meat eating is also part of the national identity. The cowboys conquering the frontier, and the settlers painstakingly moving their cattle west—that's about beef, too. Give up meat, and the cowboy dream can no longer be 100 percent yours.

If you read any contemporary book on the sociology of food, there is a pretty good chance that in a chapter dealing with meat one work will be quoted: *Meat: A Natural Symbol* by Nick Fiddes. Fiddes, a Scottish anthropologist who gave up academia to trade in kilts and tartans, put forward another theory why we find meat so tempting: it symbolizes our power over nature. To chew and to swallow other highly evolved organisms, ones that can feel and fight and bleed, is to show our human superiority. We can kill you. We can eat you. The stronger the opponent, the more prestige in depriving him of life (hunting lions in Africa is prestigious; harvesting cabbage in a field—not so much). Fiddes argues—and many sociologists agree with him—that we value meat not in spite of hurting other creatures but precisely *because* it involves hurting other creatures. If carrots suffered more when killed and fought for their lives with a bit more might, maybe vegetarian diets would carry higher status than they do now. As it is, only butchering animals can prove to the rest of nature what powerful creatures we humans are. The true kings of the jungle.

With such potent symbolism behind it, no wonder we stay hooked on meat. We humans like power, and that's precisely what meat stands for. Because it is dangerous to kill, hard to obtain, and expensive, animal flesh has come to represent power over women, over the poor, over nature, over other nations. From the voodoo practitioners of Benin to University of Pennsylvania students, we believe that by ingesting meat we somehow absorb the properties of the animal. To stop eating flesh means to risk becoming veggie-like, as fast as a cabbage and as mighty as a head of lettuce (you are what you eat, after all). If you are a man, giving up steaks could mean giving up on a patriarchal society with rich guys at the top and impoverished women at the bottom. It could mean you would become less masculine, no longer one of the "real men." And recent scientific studies confirm that those of us who hold authoritarian beliefs, who think social hierarchy is important, who seek wealth and power and support human dominance over nature, eat more meat than those who stand against inequality.

But even if meat wasn't marinated in all that powerful symbolism, it would still be hard to give up because our food habits get perpetuated generation after generation—and often without much thought. We learn our culinary likes and dislikes in the wombs of our mothers and, later, on their breasts. As children, we observe the people around us and see the pleasure of consuming animals reflected in their faces, and we learn that meat is good. Really, really good. Acquired early and reinforced daily by the culture in which we live, that's a difficult lesson to unlearn, so difficult, in fact, that over the centuries scores of vegetarian leaders have failed to convince the masses to follow their meatless ways. Still, had the winds of history blown just right, the most vocal of them may have succeeded—if only they had been a bit less radical and a bit less eccentric and had shown better appreciation of the culinary arts.

8

The Half-Crazed, Sour-Visaged Infidels, or Why Vegetarianism Failed in the Past

Over two thousand years ago, there was a man who could walk on water and heal the sick. He was a man of inner serenity and great wisdom; he was even said to have died and then reincarnated. His name was Pythagoras. Kids today learn about Pythagoras in school because of his theorem on right-angled triangles: you may still recall the equation $a^2 + b^2 = c^2$. Pythagoras was also the first to suggest that Earth is round and that the light of the moon is reflected. But there was more to his life's work than math and astronomy—although walking on water was likely not among his real achievements, just the stuff of legends.

People said Pythagoras looked striking: He was very tall and handsome. "God-like," some said. There was even a rumor that he was actually the son of Apollo and the grandson of Zeus himself. What also made him stand out was the way he dressed: he wore white robes and pants, an unusual style, since practically no one in Greece of the sixth century BCE dressed in trousers. Yet his looks and his choice of fashion were not the reason why he became something of an outsider and a laughingstock for many comedy writers. The reason—or at least one of them—was his diet.

If you lived in Paris circa 1650 or in London in the 1830s, and you decided to stop eating meat, you wouldn't tell your friends you were going vegetarian. You would probably tell them you were going Pythagorean. Until the word *vegetarian* got coined in the nineteenth century, it was Pythagoras's name that was used to describe a diet that excluded animal flesh.

Pythagoras believed in metempsychosis, the transmigration of souls. In one lifetime you could be born a human, but in your next you could well end up as a pig and get slaughtered for bacon. According to one story, Pythagoras once stopped a man beating a dog because he was convinced that in the yelps of the animal he recognized the voice of a dead friend. If souls did truly migrate from humans to animals, how could anyone ever touch meat? What if the steak on your plate was made of your great-grandmother? To avoid such risks, Pythagoras and his disciples lived on a simple diet of bread, honey, and vegetables, a diet he also believed to be healthier than a meat-based one (as modern science shows, he was probably right). For Pythagoras, as for most vegetarians until quite recently, going off meat had little to do with animal welfare. It was not about them, the other creatures. It was all about us, humans, and how being cruel impacts our psyche.

As smart as he was, Pythagoras didn't come up with his dietary ideas all by himself. He was quite likely influenced by the priests of ancient Egypt, where the concept of voluntary rejection of meat was already known five thousand years ago. There might also have been some exchange of thoughts between Pythagoras and his famed contemporaries: Buddha and Mahavira (the reformer of Jainism). It appears as too much of a coincidence that the lives of these three great philosophers overlapped and that their teachings were so in tune. But even though they all believed in the transmigration of souls and preached abstention from animal flesh, Buddha and Mahavira managed to change Asia, while Pythagoras and his students remained the subjects of ridicule.

So why has meat eating endured in Greece? Did vegetarianism fail there because Pythagoras was not associated with a religion like Buddhism or Jainism? Maybe. It also likely failed because in ancient Greece meat was usually consumed at public festivals that cemented the society and saying no to sacrificial flesh made Pythagoreans outcasts: to reject meat was to reject the whole system of the polis. Meat eating also likely endured in Greece because there was no powerful vegetarian emperor there who would support the meatless movement the way India's famed ruler, Asoka, supported the teachings of Buddha. What's more, in the times of Pythagoras, meat was prized in Greece as the food to fuel Herculean muscles and boost the performance of beloved athletes—some of whom were quite carnivorous. The wrestler Milo of Croton, for example, was famed for consuming as much as twenty pounds of meat per day. The ancient Greeks, just like Paul Rozin's students at the University of Pennsylvania, believed that "you are what you eat." They thought that consuming the flesh of a nightingale was a recipe for insomnia and would likely conclude that eating boars would make an athlete strong. But what was probably of particular importance to the Greeks' ongoing love affair with meat was that the vegetarian foods of the ancient Mediterranean were not as tempting to the senses as those served in India, with all their spices, vegetables, and fruits. Followers of Pythagoras were known to subsist on little but bread, water, and a dash of wine, while in India vegetables stewed with spices were served on scented rice, followed by dishes of flavored curd, saffron caramel, and sweet cakes with pomegranates and mangoes.

Despite Pythagoras's teachings, meat eating prevailed in Greece, and for the rest of antiquity, vegetarianism in Europe was but an elitist philosophy, a domain of outsiders. In the Rome of gladiators, vegetarianism was for radicals, for people who rejected the status quo. If you wanted to stay out of trouble, it was better to hide your veggie ideology behind a slab of meat on your plate. That's what Seneca did, and the poet Ovid. Just to be safe.

Meanwhile, in the southeast corner of the Mediterranean, another seed of vegetarian ideology started to take root. Had it succeeded, our Western food culture might have been quite different than it is now.

It's only April when I'm standing on top of Mount Nebo, a 2,680-foot mountain in central Jordan, but the air is so piercingly hot I can practically feel my skin turning into parchment. All I can see around me are barren, brownish hills with just a smudge of vegetation here and there. I was hoping for great views, but I'm not lucky: the air is not clear enough for me to spot anything past the calm mirror of the Dead Sea below, shimmering on the horizon. But when Moses stood at this same spot thousands of years earlier, ahead of him was a perfect panorama of the Valley of Jericho—the Promised Land he admired from this point for the very first time. What he also saw, and I could have seen if it wasn't for the weather, was a site where humanity's relationship with meat could have changed profoundly: Qumran. It was called the City of Salt back in biblical times, but it doesn't look like much anymore: just a sand-swept archaeological site, little more than ruins in shades of tan. Yet back in its heyday, sometime between the second half of the second century BCE and 68 CE, Qumran was a vibrant settlement that had several ritual baths, a library, a communal center, and a sophisticated water system. According to some biblical researchers, it was also the place from which a vegetarian version of Christianity was readying to spread around the world.

For those who believe in the Bible's Old Testament, the history of vegetarianism is quite straightforward. It doesn't begin in ancient Egypt or in Pythagorean Greece but right at the birth of humankind. After all, it was in the Garden of Eden, among many trees "pleasing to the eye," that the world's two very first vegetarians lived—Adam and Eve.

If ever one book and one movement could have unhooked the Western world from meat, it was the Bible and Christianity. As it was, both did the exact opposite and cemented our meat-eating ways for good. The Bible itself is either pro-veg or pro-meat, depending on whom you ask. Both sides seem to agree, though, that at first the world according to the Old Testament was indeed vegetarian, as suggested by the famed Genesis quote: "I give you every seed-bearing plant on the face of the whole earth and every tree that has fruit with seed in it. They will be yours for food." Meat eating came later, after the flood, as a kind of concession from God to his naughty kids: OK, OK, there you go, if you really want, you can eat meat. Or in the actual words of the Bible: "Everything that lives and moves about will be food for you. Just as I gave you the green plants, I now give you everything." It's a permission granted to humans living in the fallen world of sin, in an imperfect world—and that is why some philosophers interpret it as proof that, for the God of the Old Testament, vegetarianism is the right way to go and meat eating is just a temporary solution until people will again be ready for the purer state of meatlessness. In past centuries, such an interpretation of the Bible's message caused scores of people to be burned at the stake. But the history of Christianity could have developed differently, and instead of fighting vegetarianism, the Christian religion could have embraced it.

In the spring of 1947, in the desert close to Mount Nebo, a few young Bedouin shepherds were looking for their stray goat when they came across a hidden cave. Inside, they discovered a real treasure: several jars filled with fragile pieces of parchment and papyrus. What the shepherds stumbled upon were the remains of Qumran.

Some of the ancient scrolls found in the Qumran caves tell stories from the times of Jesus that didn't make it into the canon of scripture. And if the biblical scholar Robert Eisenman has it right, a few of those stories may shed a new light on the history of vegetarianism. According to Eisenman, the Qumran scrolls suggest that James the Just, known also as "the brother of Jesus" (possibly a biological one), strictly abstained from meat and was the leader of a vegetarian church community in Qumran and Jerusalem. Both Christianity and our diets would likely be different today if the followers of James and not the followers of Paul the Apostle had won the leadership of the early church. James's disciples claimed that Jesus was not a god, while those of Paul the Apostle believed in Christ's divinity and worshipped him. That, of course, was a major bone of contention and was among the reasons why Paul treated James as an archrival and fought him every way he could. Meat eating became another disputed point between the two (to Paul vegetarianism was simply a "weakness"). It was Paul and his followers who won the conflict between the early church sects, and Paul's ideas, including those about eating meat, prevailed. In 68 CE when the Roman army went to quench the Great Revolt of the Jews in Jerusalem, they destroyed the Qumran settlement for supporting the rebels. It was game over for the vegetarian sect. With time, James got written out of the Christian dogma, marginalized, and forgotten—and so did his vegetarian ideas.

Yet it's possible to envision a world in which it's not Paul the Apostle who manages to secure the biggest following but James the Just, a world in which Christianity, instead of embracing meat eating, forbids it. But that is not how history played out, of course.

Instead, meat eating became an important part of Christian orthodoxy, and vegetarianism a sign of heresy, to the point that a pale face was thought to indicate a flesh-abstaining apostate. Since so many of the heretical sects in medieval Europe preached plant-only diets, consuming animal flesh was seen as a sign of piety. In a way, the vegetarian ideas of the orthodox church's enemies fixed the followers of the church even firmer in their meat-eating ways. In order to separate the "correct" church from the dissidents, the Fourth Lateran Council in 1215 went so far as to declare that during communion believers were consuming the actual flesh of Christ, and not just a wafer-shaped symbol of it. A Catholic communion became a way to oppose heretical sects and vegetarianism itself. Today, someone who is a strict vegetarian and also a strict Catholic is put between a rock and a hard place at Sunday mass: eat the host and break your vegetarian beliefs (it's meat, after all), or decline it and stand against religion.

Meanwhile, back in the Middle Ages, refusing to eat meat may have gotten a vegetarian in real trouble. In the fourth century CE, Timothy, patriarch of Alexandria, conducted tests among the clergy: those who wouldn't swallow animal flesh were suspected to be heretical Manicheans (known also as vegetarian demon worshippers). In France, hundreds of Cathars were burnt at the stake for believing that meat was a product of the devil. The same thing happened to the vegetarian Bogomils in Istanbul. Of course, both these groups didn't die solely because of their refusal to indulge in meat. Abstaining from animal flesh was just a sign that their religion and politics were on the wrong side. The Bogomils were a thorn in the side of the Byzantine rulers because they preached the liberation of the Slavs. The Cathars died because they were caught up in the war between northern and southern France.

It was bad for vegetarianism that it got attached to politics and religion and, what's worse, to the weaker, losing camps. On the flip side, during the Middle Ages the masses would not have followed a vegetarian ideology without it being strongly backed by religion. In medieval Europe, if you had a chance to eat meat, you'd be a fool to refuse it. There was rarely any food to waste, particularly food so nutritious and calorie dense as meat. Europe wasn't India, with its abundance of beans and lentils, which are great meat substitutes. A vegetarian religion, either vegetarian Christianity à la James the Just or based on one of the so-called heresies, could have made eating meat sinful, morally aversive, and this might have been enough to keep starved peasants away from animal flesh. As it was, meat eating endured because those who encouraged it proved more powerful.

Yet mighty enemies likely weren't the sole reason why vegetarianism didn't catch on in the Middle Ages. The heretical doctrines may have lost some support from the poor by being simply too radical and too ascetic. As Robert Eisenman once told me, the connection between vegetarianism and purity goes back all the way to the community of Qumran. For them, avoiding meat was about "not polluting the Heavenly Host with unclean foods" (in Eisenman's words). A similar thread comes up later on in Europe. The medieval heretics not only claimed meat eating to be a sin, but they also believed in more far-reaching purity: they preached total sexual abstinence, and they forbade alcohol. In other words, they weren't much fun. That's a theme in the history of vegetarianism that reappears over and over. Even in mainstream church doctrine, not eating meat was tantamount to mortification of the body, a bit like wearing a cilice. You did it during official fasts; you did it if you were a monk. Temporary vegetarianism was thought to decrease the flow of semen and control lust. Again: not fun.

Catholics and heretics would probably also agree on another issue: abstaining from meat had little to do with the suffering of animals. Just as for Pythagoras, vegetarianism was about the human soul, not about cows and pigs. God clearly gave animals for humans to rule. Besides, animals were nothing like humans, who were created in the image of God. That's what St. Francis the meat-eating animal lover believed. That's what St. Thomas Aquinas believed, too; he even said: "It matters not how man behaves to animals." And to St. Augustine not killing animals was merely a superstition. This permission to think of animals as inferior, granted by the Bible and ratified by church philosophers, boosted our appetites for meat until the dawn of the industrial era. It took a long time, until the mid-nineteenth century, for humanity to start considering animal suffering as a reason to stop killing them for food.

John Harvey Kellogg was a very short man, barely five feet four inches. He claimed only his legs were short and preferred to receive visitors when seated, but he made up for a small frame with a big ego. He was controlling, imperious, extremely ambitious, and ridiculously hardworking—he often clocked twenty-hour workdays. His goal was to improve the health of Americans by changing their daily habits. He always dressed in white, from head to toe. He believed that clothing should be comfortable and opposed impermeable fabrics, corsets, and high heels. Inspired by contemporary writers who advocated the health benefits of plant-based diets, he gave up meat and went vegetarian. What's more, he believed animals to be "sentient creatures" who deserved to be treated not as "sticks or stones, . . . but *beings*" and not killed for food.

Kellogg was a medical doctor and a skilled surgeon, and his training and reputation helped him convert others to vegetarian diets. During one of his popular lectures, he placed a steak under one microscope and a pile of manure under another and, after comparing them, declared that there were more bacteria in the meat. In 1876 he became a superintendent of Western Health Reform Institute in Battle Creek, which he renamed Sanitarium and turned into "the" place to go for the wealthy and famous of the day who wanted to improve their health. At the height of its popularity, the heavy, imposing building of the "San" saw seven thousand guests per year, including Henry Ford, J. C. Penney, Amelia Earhart, and Thomas Edison. The diet offered at Battle Creek Sanitarium was soon made completely vegetarian and, admittedly, rather bland: zero spices, low fat, and low protein. Thankfully, though, products developed under the guidance of Kellogg's wife, Ella, in the Sanitarium's test kitchen proved more appealing to the senses: granola, soy milk, and peanut butter, which at the very beginning were destined for patients who couldn't chew up nuts well (Kellogg claimed proper mastication to be very important for health). Then, of course, there were the cornflakes—flattened and baked maize kernels that became such a tremendous success.

With his outlandishness, Kellogg fit well with the rest of the vegetarian leaders in the nineteenth-century UK and US. They were a truly colorful bunch: radical, vocal, often naïve, and eccentric. Though they may have failed to turn their fellow Americans and Britons into plant eaters for good, they didn't fail completely. In a way, they had a great impact on the way we eat and turned the majority of us into part-time vegetarians.

Think about it: What's the so-called traditional British or American breakfast food? Eggs, bacon, and sausage. In other words—heaps of meat. And yet what do most of us actually eat to start the day? Cereal, PB&J, pastries. Less than 20 percent of Americans have meat for breakfast, and by far the most common morning staple is cold cereal, with peanut butter sandwiches being another big favorite. If you like these foods and eat them often, you should thank nineteenth-century vegetarians and in particular John Harvey Kellogg and his wife, Ella. But the Kelloggs weren't the only ones who developed vegetarian substitutes that went mainstream. There was another line of products promoted for meatless living that took America by storm: graham bread, graham crackers, and graham flour.

Sylvester Graham wasn't a doctor. He wasn't formally trained in medicine, just self-taught. And yet he propagated some radical ideas in health care, such as exercise, fresh air, and frequent baths. The youngest of seventeen children of an elderly minister and a mentally ill woman, Graham was a sickly child who nevertheless grew up to be, as some claim, "the father of American vegetarianism." Or, as Ralph Waldo Emerson said, the "Prophet of bran bread and pumpkins." Although Graham didn't invent whole-grain breads and flours, he did a lot to promote them, believing them to be a healthier option than white, refined products. Graham also believed in the healthiness of plant-based diets and worked hard to popularize them. Just as early vegetarians were called Pythagoreans, in the nineteenth-century US they became known as Grahamites. In 1832 the movement was given an unusual boost: a cholera epidemic. Most Americans believed that to keep from getting sick, one had to eat tons of meat and avoid veggies and fruits as evil incarnate—they were thought to cause disease. Even governments supported such claims: for example, the Board of Health in Washington, DC, advised that potatoes, beets, tomatoes, and onions should be eaten in "moderation."

And yet, the beet avoiders were still getting sick and dying. Graham, meanwhile, took another stance. Besides advocating baths and fresh air, he told Americans to eat the feared veggies and forget the meat. Since he was a skilled orator, his lectures attracted thousands. And since the meat-based diet wasn't working against cholera, some people were ready to try the vegetarian one. Many Grahamites claimed they started feeling better on the new regime, and vegetarian ranks started to swell.

The mid-nineteenth century was a pretty good time for vegetarians. Not only were they no longer burned at the stake (which is always a plus) but also the first nonreligious vegetarian organizations were starting to form. This relative boom in the vegetarian movement wouldn't have been possible if it wasn't for the fact that meat became more easily available across western Europe and the US. More vegetables, cereals, and meat substitutes showed up in stores, too, so it became easier to base your diet on plants. The center of the vegetarian universe moved to Great Britain, a country where the meatless Indian cuisine was well-known, where urbanization put people and their ideas closer together and made them long for nature, where the Protestant religion afforded more freedom to contemplate animal rights than Catholicism ever did, and where Charles Darwin preached that animals and humans were not so different after all.

In 1847, in a simple but stately Northwood Villa in Ramsgate, UK, a group of Pythagoreans set up a meeting. Some had been off meat for as long as thirty-eight years, and yet, as one contemporary journal reported: "They all looked truly patriarchal, healthy, strong, and full of intelligence and love." They arrived in Ramsgate to form a new organization, the Vegetarian Society, the first of its kind in the world. That day over 150 people committed to a "fruit and farinaceous diet" (*farinaceous* meaning "rich in starch"). It was also then that the term *vegetarian* was officially adopted, from the Latin *vegetus*, which means "one who lives a healthy life." Soon similar organizations sprouted up in other countries—in 1850 in the US, in 1886 in Australia.

If "fruit and farinaceous" doesn't sound very appetizing to you, that's because it wasn't. Vegetarian food was one of the main reasons why the nineteenth-century vegetarian movement didn't attract the masses. Simply put, it was bad: bland, overcooked, unseasoned. If you went to a vegetarian restaurant circa 1890 or even 1920, you wouldn't get "portabella carpaccio with pickled eggplant and caper purée," a dish currently offered in Vedge, a vegetarian restaurant in Philadelphia. You would get "pale, languorous carrots, fraternized with pecks of weepy boiled beets, and . . . at least a silo full of various greens and grasses," as one journalist reported. Spices, and even salt, were considered as harmful as alcohol and religiously avoided. Vegetables were often boiled into mushiness—one writer actually recommended that all veggies be so thoroughly cooked as to "prevent any crispness." Even John Harvey Kellogg subsisted on little but apples, graham crackers, potatoes, and oatmeal gruel. To upper, aristocratic classes, with their taste buds trained on gourmet feasts, vegetarian food must have been impossibly boring. To lower, aspiring classes, it offered nothing to aspire to.

And if bad vegetarian cuisine wasn't enough to keep people eating meat, the nineteenth-century vegetarian movement scared some potential followers with its asceticism—just as the Pythagorean and the heretical thoughts did before it. Grahamites wanted to fight many evils. Meat was evil. Tobacco was evil. Alcohol was evil. Sex was evil. Graham lectured that sexual intercourse overstimulated the body and made it prone to diseases. Kellogg went even further, calling for female circumcision so that women wouldn't derive any pleasure from sex. The leaders of the movement gave personal examples of austerity, too. Kellogg slept on the floor with newspapers for a mattress, while William Alcott (an American promoter of vegetarianism) started his days at 4:00 a.m. with a cold-water bath. Meanwhile, Leo Tolstoy (an on-and-off vegetarian) called for the rich to give away their money and land in order to go back to nature. You can guess Grahamites didn't win many followers with that, either.

The vegetarians didn't only struggle with the image of being excessively puritanical. They were also often considered too radical, too naïve, or, simply speaking, wacko. They were called "half-crazed," "sour-visaged," "infidels," and "food cranks." And, admittedly, some of them had pretty unusual ideas. Over one hundred years before the first hippie communes, Amos Bronson Alcott (the father of *Little Women*'s Louisa May) started a vegan community in New England, which he named Fruitlands. Alcott actually believed that he was planting the seeds for a garden of Eden version 2.0 and that Harvard, Massachusetts, was the promised land. Everyone in Fruitlands wore specially designed tunics and lived on plants grown in the commune's soil—which, by the way, remained unfertilized, because Alcott claimed manure was too filthy. That proved costly, particularly combined with the inhabitants' lack of agricultural skills. Once the first winter arrived, Fruitlands went bankrupt: the crops had failed, and there was nothing to eat.

But at least Fruitlands did exist for a couple of months. A vegetarian metropolis in Kansas called Octagon City never really took off. The plans were grand: the city was to be sixteen square miles, with an agricultural college and scientific institute, and to support itself by exporting fruits, graham flour, and graham crackers. Yet before Octagon City managed to start exporting anything, the first settlers were chased away by snakes, outlaws, hostile Indians, and . . . mosquitoes.

Such failing, far-out ideas as Fruitlands and Octagon City would have been reason enough to keep the masses from joining the vegetarian movement. But fate handed the critics of vegetarianism an easy weapon: the premature deaths of several of the movement's leaders. First, Graham died at the age of fifty-seven. Then William Alcott expired at age sixty. These deaths were followed by those of James Simpson, president of the Vegetarian Society, at age forty-eight, and Anna Kingsford, famed British vegetarian physician, at forty-two. Even though these deaths had little to do with the absence of dietary meat and were the results of tuberculosis, preexisting medical conditions, or poisoning with industrial fumes, the word spread. People started to wonder: maybe vegetarian living wasn't so healthy after all.

And then came the wars. Despite the nineteenth-century vegetarian movement suffering from some bad press, it was still relatively strong. After all, it did manage to convince many Americans and Britons to remove bacon and sausage from their breakfasts. But the two world wars pushed Western diets firmly back toward meat. To begin with, it's hard to care for animals when you see so much human suffering around. As one novelist wrote: "These days nobody wept for horses." Nobody wept for chickens and pigs, either. Besides, if you were an American or a British soldier, you could hardly be a vegetarian. Army rations were heavy on meat, so if you wanted your stomach full, you had little choice but to eat everything up, which most did with pleasure, of course. For the many poor who swelled the military ranks, the newfound abundance of animal protein was a dream come true. They could finally, often for the first time in their lives, eat as much meat as they wanted, meat that for so long had symbolized unreachable power and luxury and that was loaded with umami, fat, and the flavors of the Maillard reaction.

Meanwhile, for American and European civilians, meat was such a rare treat during the wars that it became even more of a status symbol than before. Social psychologists would tell you this was an example of the scarcity principle at work: the less available something is, the more we value it. When a 1940 survey was conducted to find out what Americans craved to eat the most, first place was taken by ham and eggs, followed by prime ribs, chicken, lobster, and baked Virginia ham. No vegetarian dishes made the list.

At the same time, something strange was happening in UK: the numbers of vegetarians-by-choice actually went up during World War II. Did the hardships of war make the British link animal suffering with the human one and inspire them to stop eating meat? Not exactly. What really happened was far more prosaic. In Britain, if you registered as a vegetarian, you were allotted a bigger ration of cheese, much more attractive than the tiny and unreliable meat ration. If you wanted to feed your family, the choice seemed obvious. Yet as soon as pork and beef became available again after the war, all these "vegetarians" were more than happy to sink their knives and forks into meat—probably even more so than before, since the new postwar abundance of meat became a symbol of peace and prosperity.

War scarcity, human suffering that left no place for caring about animals, meat-loaded army rations—that would have been enough to stamp out the vegetarian movement. But it also didn't help the movement's PR that Hitler was a vegetarian. Several of the dictator's biographers wrote about his commitment to a plant-based diet. He was quite a hypochondriac: he mistook stomach cramps for cancer and worried about his muscles trembling. A vegetarian diet was supposed to help preserve his precious health. At the same time, in a move that seems bizarre to many, Hitler outlawed vegetarian societies in the Reich. It wasn't bizarre at all, though. Not only did the Führer want to set himself apart from the plant-munching weirdos, he also didn't like the radical, counterculture currents that were part of the vegetarian movement.

As years passed and war scarcities became a distant memory, as new, more scientifically sound studies started showing the benefits of meatless diets, the numbers of vegetarians in the West inched up. Still, just like the Grahamites in the nineteenth century or the Pythagoreans in fifth century BCE, these modern plant eaters were considered outsiders who dared to reject societal norms. As *Rags* magazine reported in 1971: "To many Americans, vegetarianism represents another weirdo protest of the head generation against mom-and-apple-pie-ism." But something *was* different in the '60s and '70s: being a weirdo wasn't so bad anymore. In the era of entertainment media and television, granola-loving hippies were news. They grabbed attention and became a group celebrity. But when the '70s waned into the '80s, and consumers tuned their TVs to shows such as *Dallas* and *Dynasty*, depicting upscale lifestyles, hippie values lost out to materialism. Whatever symbolized power and strength was good and desirable, whether it was steaks or Patek Philippe watches.

So far Western vegetarians hadn't managed to unhook humanity from meat—not the hippies, not the Pythagoreans, not the Grahamites. A few reasons for this failure have reappeared, refrain-like, throughout history. They were too radical, these vegetarians, too eager to reject mainstream values. They were too ascetic and puritannical, turning down not only meat but other sensory pleasures as well—sex, alcohol, tobacco. They lacked the support of the mighty of the day. They got mixed up in politics on the wrong side and lost. Maybe the West would have been far more vegetarian if Pythagoras had lived not in democratic Greece but in Rome and had managed to convince a powerful emperor to follow his diet, an emperor who could have done as much to spread of vegetarianism in Europe as Asoka did in India. Maybe the West would be vegetarian if James the brother of Jesus had led Christianity instead of Paul the Apostle and if the Romans hadn't wiped out the Qumran community for joining the Great Revolt. In the past it was religion that had the power to turn nations to plant-based diets. Vegetarians could not have won by preaching animal rights: there was too much human misery for the masses to care about the suffering of animals.

But maybe vegetarianism was never likely to succeed in Europe for one simple reason: the lack of meat replacements. Europe was no India with its abundance of protein-rich pulses, fatty oilseeds, and spices for flavoring vegetable dishes. For the poor, vegetarianism was hard to follow: when you have little to eat, you don't say no to a roasted duck. Meanwhile, the elites, who didn't suffer from the empty-plate syndrome, weren't eager to sacrifice their culinary pleasures. European plant cooking was mouth-numbingly bland. For those who put good eating on a pedestal, overcooked carrots and wilted greens don't hold much appeal. This martyrdom of taste buds championed by vegetarians from Pythagoras to Graham might have been a mistake. Maybe if the rich could have eaten gourmet foods without meat, more would have supported the meatless movements. As it was, meat eating prevailed. The long-held symbolism of meat, powered by its scarcity and combined with the umami and fat of its flavor, and with our protein hunger, proved stronger than the arguments of vegetarian philosophers.

You would think that it's much easier for modern, twenty-first-century vegetarians to convince others to follow their path—but that's not necessarily the case, either. Not only are there genetic reasons for why some of us find it harder to give up meat and try novel diets (such as a plant-based one), but also basic human psychology often stands in the way of vegetarian arguments—and keeps us hooked on meat.

9

Why Giving Up Meat May Be Harder for Some of Us

The air in Boston's Reggie Lewis Athletic Center is full of the aroma of meat. "Smells pretty good in here," says a woman dressed in a floor-length, flowery skirt, as she enters an ultracrowded hall. Hundreds of people are pressed close together, slowly pushing toward the myriad stands ahead. Something delicious is being cooked inside, that's for sure. As we approach, the scent grows stronger, and I can make out sausages, crispy bacon, juicy burgers, all being cooked fresh.

Yet there is no meat in sight.

The event to which the crowds have swarmed is the annual Boston Vegetarian Food Festival. Everything edible here is made of plants—not even milk or eggs are allowed to cross the threshold. The hall is, of course, packed with vegans and vegetarians. And yet the woman in the hippie-like flowery skirt, perhaps more obviously identifiable as a vegetarian, is an exception. You would never have guessed by looking at most of the people at the center that they are plant eaters attending a vegetarian festival. Some are young, some are old; some are thin, some are fat. Some wear sweatpants, and others wear suede suit jackets with elbow patches.

Surveys have shown that many people think vegetarians are somehow different from the rest of society. To the average American, the folks who turn down meat are pale, pacifist, hypochondriacal, foreign-car-driving liberals. Thus the question becomes: Who *are* modern vegetarians? Are they born to be plant eaters? Is there something in their genes or personality that makes it easier for them to stop eating meat? Can you tell someone is a vegetarian by scanning his or her brain? And why on Earth do some of them eat fake meat "sausages" and "burgers" and no-fish "tunas"? Are they fooling themselves, secretly craving meat after all?

An American comedian, David Brenner, once said that "a vegetarian is a person who won't eat anything that can have children." If only it were so simple. There is no clear-cut definition of a vegetarian on which everyone agrees. Even vegetarians disagree about what constitutes a vegetarian diet. Pescaterians eat no meat but do allow for fish. Others won't eat fish but think that devouring a plate of mussels is perfectly OK. To add to the confusion, there are pesco pollo vegetarians who avoid red meat but eat chicken and fish, flexitarians who generally avoid meat but still eat it on occasion, and the VB6 folks—who eat only vegan before 6 p.m. but after that hour may dig into steaks. Meanwhile, orthodox vegans won't even touch honey because it comes from the exploitation of bees.

No wonder it's hard to tell how many vegetarians are truly out there. Recent estimates show that 3 to 5 percent of Americans consider themselves vegetarian, just like 4 to 8 percent of Canadians, 3 percent of Australians, 2 to 5 percent of British, and a mere 0.3 percent of Portuguese, who appear to be the most meathooked of Western nations. The discrepancies in the statistics come partly from the fact that people who are flexitarians may sometimes say they are vegetarian and sometimes admit they really aren't—mainly depending on how survey questions are designed. Most studies show that in Western countries, the US included, the numbers of vegetarians are not particularly high. Che Green, executive director of the Harris Interactive Service Bureau, went so far as to call vegans and vegetarians "a blip on the demographic radar." He said: "Statistically speaking, we're below the margin of error for most surveys."

And yet, one can't help but feel that vegetarianism is on the rise. It seems that wherever you look, there is a vegetarian celebrity: Natalie

Portman, Anne Hathaway, Liv Tyler, Pamela Anderson, Moby, Alanis Morissette, Bill Clinton, Chelsea Clinton, Dustin Hoffman—the list goes on. In Europe, the world's first vegan supermarket chain, Veganz, is opening one new location after another, and similar stores are popping up all over North America. Flip through a newspaper, turn on the TV, or scroll through the trending topics your friends are posting to social media, and soon enough you'll find even more hints that folks may be leaving the ranks of meat eaters in droves.

The most common reasons people cite for why they go "veg" are health and animal welfare. Although, admittedly, some people may have more obscure motivations. As *Saturday Night Live* comedian A. Whitney Brown once joked: "I am not a vegetarian because I love animals. I am a vegetarian because I hate plants." Health vegetarians usually change their diets slowly, eliminating meats one by one, starting with red, then ditching poultry, and finally dropping fish. Those who stop eating meat for animal reasons often do so more suddenly. They may watch an undercover slaughterhouse video or see an animal killed in front of them and immediately decide to stop eating all meat—scientists call it a "conversion experience."

Psychologists say that it's easier to cut meat out of your diet when you are going through big life changes such as divorce or moving to college. Giving up beef or pork is a lot about rejecting identities. Sharing food solidifies social bonds and makes people feel they belong. If you are Chinese and stop eating rice, for example, you are casting off a large part of your ethnic background. The same goes for saying no to an American Thanksgiving turkey. While moving to a new city, or a new country, it's easier to shed not only your previous geographically identified self but also your meat-eating self. Easier—yet rarely easy, since vegetarian clichés can discourage many from joining the ranks of the meatless.

When I meet Kate Jacoby at the Vedge restaurant in Philadelphia, a vegetarian-food haven she runs with her husband, chef Richard Landau, she doesn't strike me as a stereotypical vegan. Dressed in neat slacks and a tailored cream top that complements her light hair, she gives my hand a good shake as she lets me into the empty restaurant, a few hours before opening. Neither does her restaurant conform to vegetarian clichés. Vedge is elegant with a modern twist. Resolutely high-end, there is no hippie stuff anywhere in sight. To my disappointment, the air is devoid of any of the delicious scents that usually fill this place, no savory smells of grilling, no balmy aromas of sauces. Nothing is being cooked on the stoves of Vedge, not yet. The only thing that flows toward me from the kitchen is a rhythmic, unrelenting clack, clack, clack: the sound of countless vegetables being chopped.

Jacoby has been off meat most of her adult life. When I ask her about the clichés surrounding vegetarians, she sighs and says: "There is this need for vegetarians to be perfect. They can never get sick because people will say: 'Oh, that's your diet.' Your vegan shoes can never rip because they'll say: 'Oh, these are vegan; these are inferior.'" Jacoby and her husband take pains to take care of themselves to put a good face on the meat-free movement. "We want to have nice hair and look good, set an example," she tells me, then adds: "I see a lot of animal rights people who take extra steps to make sure they don't come across as weaklings, and many vegetarian athletes who work out to show people that they can have muscles and be fit."

For many omnivores, vegetarian clichés can reinforce their commitment to meat-based diets. The author of the recently published *The Shameless Carnivore: A Manifesto for Meat Lovers* admits, for example, that some meat eaters, including himself, call vegetarians "soy-heads, veggieburgers, communists, the enemy." In the past, even scientists were not immune to the belief that there was something inherently wrong with vegetarians. Back in the 1940s, the head psychiatrist of one Long Island hospital posited that vegetarians are secretly sadistic and "display little regard for the suffering of their fellow human beings." Twenty-first-century studies prove that this is simply not true: fMRI scans reveal that if you show vegetarians and omnivores pictures of human suffering, the empathy-related areas of the brain will activate more in the plant-eating group.

Nevertheless, the clichés that surround vegetarians are quite pervasive, even today. In one experiment, researchers presented volunteers with five types of diets, including a vegetarian one, a gourmet one, and a fast-food one, and asked them to describe the personalities of the people who consume such fare—in the spirit of the French writer Brillat-Savarin who famously observed: "Tell me what you eat and I will tell you what you are." The results were far from surprising. Fast-food types were believed to be religious, conservative, and fond of polyester clothing. Gourmets were liberal and sophisticated. Vegetarians were foreign-car-driving pacifists. In other studies, meat eaters described vegetarians as generally good people but also weak, weight conscious, and pro-drugs. Another cliché that got stuck to the image of vegetarian men in particular is that of a sex-deprived wimp. Psychologists have found that not only omnivores but even *vegetarians* think of vegetarians as less masculine. Such notions, of course, stem from the strong connection among meat, blood, power, and manliness that is inscribed in our culture. This perceived lack of masculinity is likely one of the reasons why 30 percent of omnivores say they wouldn't date a vegetarian. But meat avoiders are often not much into meat eaters, either. In 2007 a new word made headlines in the Anglophone world: "vegansexuality." The source: a study in New Zealand that showed that vegan women didn't like meat-eating partners as much as they did plant-eating ones. The word was coined to describe vegans who prefer only to have sex with other vegans. All of a sudden, vegansexuals started "coming out," which in turn resulted in more clichés being attached to them. On Internet forums, the vegansexuals were called bitter pleasure deniers and "notoriously bad lays." Such attacks may stem from vegans appearing to reject the cultural norm, which links meat eating with masculinity and sex.

Some vegansexuals claim that vegans really *are* different—that they even *smell* different. According to one experiment done in Europe, there may actually be some truth to it: female students rated the scent of men on a meat-based diet as less pleasant than the smell of the very same guys when they went vegetarian for two weeks. Yet a better way to tell if someone is a vegetarian is to look at his or her hair—not the hairstyle or color, mind you, but the hair's chemical composition. If you send a few locks to a lab, it can be determined by the abundance of ^{13}C and ^{15}N proteins whether their owner's diet is made up of plants or meat. And if you were really set on checking whether someone is a meat eater, you could also scan his or her brain. By placing electrodes on a person's scalp to measure electrical activity in the brain, and then showing that person pictures of meat, scientists can see a difference between the reactions of vegetarians and omnivores—meat stimulates a vegetarian brain more than it does an omnivore brain.

Does this mean that vegetarians are somehow fundamentally different from meat eaters, that they are born different? Or do they just become different once they stop eating meat? It appears that, to a certain degree, both these statements are true. Although the scent and hair composition of anyone will change after switching to a plant-based diet, there may be some inborn characteristics that make it easier for some people to give up steaks and bacon in the first place while others may have a genetic makeup that reinforces their desire for meat. A study conducted on British pairs of twins showed, for example, that 78 percent of how much we like meat or fish is heritable—meaning that if your parents dislike beef and pork, you are more likely to have an aversion to them, too. Another study, this time from Brazil, suggested that serotonin receptor genes 5-HT, which are involved in the development of bulimia and binge eating, are also responsible for how much some people like beef, although, to be fair, the effect was pretty small.

There is, however, one more inherited characteristic that could make switching to a completely new diet—like a vegetarian or a vegan one—a bit more difficult for some people. To survive, omnivores, such as humans, rats, and cockroaches, rely on two mechanisms that pull them in opposite directions: food neophilia (a temptation to try new things in case they may be tasty and nutritious) and food neophobia (a fear that these new foods may kill them). Some people are more neophobic than others. Twin studies show that the degree to which we fear novel foods is about two-thirds dependent on what we've inherited from our parents. Neophobes are the people who don't like to check out new restaurants, don't enjoy ethnic cuisines, and tend to dislike anything they can't easily recognize. They may also have more aversion to vegetables and fruits and so might be less eager to try, say, veggie burgers—if they haven't grown up eating them. Neophobes who do become vegetarian tend to eat more junk food or to have a more limited diet: they throw out the meat but replace it with nothing new. From "meat and potatoes" guys, they can turn into "just potatoes" guys.

Studies have linked two more inborn traits that indicate which people are more inclined to switch to a plant-based diet. The first of them is "openness to experience," one of the big five personality traits (the others being extraversion/introversion, friendliness/hostility, conscientiousness, and neuroticism/emotional stability). People high in openness generally prefer new ideas, are intellectually curious, and lean toward nontraditional values. They are also less likely to be devout meat eaters and, instead, consume more veggies and grains. The second trait connected to meat eating is IQ. Research shows that someone's IQ at age ten predicts the likelihood of that person becoming a vegetarian later in life: the higher a person's IQ, the less likely he or she will be a meat lover as an adult.

Yet these apparent differences in personality traits between vegetarians and meat eaters don't really explain why the members of these two food camps so often cross swords over dinner. But scientists do have an idea of what does.

Evelyn Kimber, president of the Boston Vegetarian Society, is quite unlike Jacoby but is nevertheless still quite different from any of the common hippie-veggie stereotypes. With her simple blouse and a necklace of enamel fruits, she makes me think of home-baked cookies and Sunday neighborhood fairs. Like Jacoby, Kimber believes that the face the vegetarian movement presents to the world is very important if it wants to encourage others to reduce their meat consumption. This face, for starters, should be peaceful and nonconfrontational. That is why at the food festival, of which she was the organizer, no vendor was allowed to display anything with slogans that seem hostile toward meat eaters. "We want meat eaters to feel welcome and our messages to be positive," she tells me. Kimber herself tries not to behave in any way that would, in her words, "put people off." That means, for example, using the phrase "I don't eat meat" in place of "I'm vegan," which tends to aggravate omnivores more. After all, the very idea of a vegetarian identity is something that can easily turn a nice dinnertime conversation into a heated argument.

Imagine this scene: In a dining room, a table is set. There are candles and linen and a vase brimming with flowers. And food, of course—tons of it. Plates of crispy salads, charred meats, fragrant sauces. There are five people seated at the table: a committed meat lover, an average meat eater, a health vegetarian, an ethical vegetarian, and a vegan. It is an explosive mix.

The meat eater reaches for a platter of chicken and offers it to the vegan. The vegan refuses. A question follows, then an answer, and soon everyone is deep into a heated discussion about the rights and wrongs of eating meat. Voices get raised. Hands gesticulate. And yet, one might wonder: Who are the two seated at the table who are most at odds? Is it the meat lover and the vegan? Not necessarily. More likely, the average meat eater, someone a bit less committed to enjoying animal protein, would turn against the vegan and the ethical vegetarian. Why? A recent experiment showed that people actually argue more fervently when they are less confident about their dietary choices.

People also tend to get more vocal if you criticize their personality rather than just their actions, as in "you are silly" versus "your behavior is silly." That's why health vegetarians are usually less touchy about their diets. Since for vegans and for ethical vegetarians (those who went "veg" for the health of the chickens, not their own—to paraphrase Isaac Bashevis Singer) not eating meat is more of a lifestyle choice than a culinary preference, they are also more likely to feel threatened by meat eaters' accusations. That is also why, in another surprising turn of events at our dinner table, the ethical vegetarian may attack the health vegetarian, whom he perceives—according to surveys—as selfish. The presence of meat eaters would make a conflict between the vegan and the vegetarians more likely, too. In experiments, putting an omnivore between vegans and vegetarians spotlights the moral issues of diets, and vegans start accusing meat-avoiding milk drinkers of being hypocritical. The result is a nasty fight.

In how-to books addressed to vegetarians, there is often some space dedicated to answering meat eaters' inquiries: Why are you wearing shoes made of leather? What do you feed your pets? Don't you have to kill vegetables to eat them, too? However, there is rarely any attempt at figuring out why meat eaters ask these questions in the first place, or why vegetarians' arguments almost never manage to get the omnivores to give up meat. What's so special about beef or pork that it can turn a dinnertime conversation into an argument? Why doesn't telling people you don't eat carrots stir up similar emotions?

Kristin Lajeunesse, an author whom I meet among the leaflet-coated stands of the Boston Vegetarian Food Festival and who quit her office job to wander across the US and write about vegan food, believes that when you tell meat lovers that you don't eat animals, what they often hear is "Oh, you think I'm a bad person because I like meat." Lajeunesse is onto something. Studies show that a mere exposure to a plant eater (as opposed to any other dieter) puts omnivores on edge and causes cognitive dissonance, turning on a set of psychological mechanisms that ends up allowing the meat eaters to double down on their carnivory.

We experience cognitive dissonance whenever our beliefs and our behaviors don't match. Say you think that driving SUVs is bad for the planet, yet you really want a Hummer—you just love how it looks and how it drives—and so you buy it. That nasty feeling you experience whenever you climb into the truck thinking about pollution is cognitive dissonance. You want to get rid of it. You can either change your behavior (sell the car), change your beliefs (that's difficult and unlikely), or—here comes the most popular option—rationalize your actions. You may tell yourself you had no other choice. That Hummers are safer. That whatever you do won't impact the climate much anyway.

People who believe that killing animals is perfectly fine may not experience cognitive dissonance about eating meat. But those omnivores who would prefer if the beef or pork on their plates was 100 percent cruelty free (which is impossible unless the meat is lab grown) need to apply what psychologists call "dissonance-reducing strategies" in order to avoid unpleasant feelings—and to be able to maintain their diets. A common strategy, "denigrating the victim," is to convince yourself that animals are not very smart and that they can't feel pain. As a journalist for Hog Farm Management once suggested: "Forget the pig is an animal. Treat him just like a machine in a factory."

To learn more about psychological mechanisms that enable us to justify eating meat, I called Brock Bastian, a researcher at the University of New South Wales, Australia. Bastian was raised a vegetarian but threw in the towel when he was an adult and started eating meat. Yet something seemed not quite right: he kept feeling guilty about his choice of diet. And that is why he set out to investigate what makes people—including himself—so uncomfortable about turning animals into food.

Over the years, Bastian has conducted a series of studies that have shown that eating meat makes people think of animals as rather dumb creatures devoid of emotions. In one of his experiments, Bastian asked the study participants to rate the extent to which a sheep or a cow possessed certain mental capabilities, such as desiring, wishing, and thinking. Later, he told the volunteers they would now take part in another unrelated "consumer behavior" study (that was not true; the studies were very related indeed). Some of the volunteers were asked to write an essay about sampling apples, while a plate of green apples was placed in front of them. Another group had to write about meat, as a dish of roast beef or lamb "infused with rosemary and garlic" was put on the table. Once they finished writing, the experimenter told them he was just going to get some cutlery so they could dig into the dish and asked them to fill out another questionnaire about cows and sheep, as they waited. Again, they had to rate the mental capabilities of the animals. The results confirmed what Bastian suspected all along: people changed their perception of cows' mental abilities compared to what they had said only minutes earlier if they thought they were just about to eat meat. In the second survey, they described the animals as less smart and less thoughtful than they did before. What's more, such belittling of animals made the meat eaters feel better. "Thinking of a cow suffering and dying so that we can eat beef makes us uncomfortable," Bastian tells me. Therefore, we convince ourselves that cows are stupid, cannot perceive much pain anyway, and cannot suffer. This helps us escape the cognitive dissonance and enjoy the roasts and steaks on our plates.

It's not just eating meat, though, that can make people lose moral concern for animals; even categorizing a species as food may suffice. To check this idea out, Bastian conducted another experiment. This time he made volunteers read slightly different versions of an article about Bennett's tree kangaroos living in the lush rain forests of Papua New Guinea. Some of the study's participants learned that tree kangaroos are killed for their meat by local tribes. Others, that these animals are never hunted and never eaten. Afterward, Bastian asked the volunteers how much a kangaroo would suffer if harmed. Those who read the version of the story presenting the animals as food judged this species as not capable of experiencing much pain—a classic dissonance-reducing strategy.

While men are more likely to denigrate animals as a dissonance-reducing technique, women often prefer not to think about the animal at all and disconnect living creatures from the food on their plates: scientists call this approach "dissociation." What helps them to continue eating meat is our language. After all, it's easier to forget about dead cows if we brand them "beef" and dead pigs if we call them "pork." The eighteenth-century Japanese went even further, renaming horsemeat "cherry," deer "maple," and wild boar "peony." In less poetic language, the modern meat industry nevertheless refers to cows and pigs as "grain-consuming animal units." Would we be so willing to keep eating meat if we called it—as George Bernard Shaw suggested—"scorched corpses of animals"? Probably not. Meanwhile, we don't have special names for carrots or cabbage. Living or dead, a carrot is still called a carrot—likely because we have no need of hiding where the food comes from since we don't consider depriving carrots of life a moral issue.

Even the sheer number of animals butchered for meat may help us feel better about eating them. Experiments show that the greater the number of victims—say, of an accident or a natural disaster—the less people are inclined to care. As little as two deaths already seem less worthy of concern than just one. Fifty-eight billion chickens slaughtered across the globe each year thus seems like a mere statistic.

Returning to our heated dinnertime discussion: according to studies, it's enough for omnivores to face vegetarians, or even just think about them, for their meat-related cognitive dissonance to kick in. It's an unpleasant feeling, so to push it away, they shift the attention to the vegetarian. By making the vegetarian appear inconsistent and morally dodgy (those leather shoes), the meat eaters can quiet their own inner conflict.

One thing that could help calm the air would be for the vegetarians and the vegan to say that they not only secretly crave meat but actually keep a stash of beef jerky under their beds and munch on it when nobody is around. This would make them less morally threatening. Nor are such meat-eating vegetarians a rare thing, it appears. In one Canadian survey, a mind-boggling 61 percent of "vegetarians" admitted to eating poultry, and 20 percent to occasional feasts of red meat. In another poll, in the US this time, 60 percent of self-proclaimed vegetarians had had some animal flesh within the last twenty-four hours. This means that the number of committed vegetarians and vegans in the US may be as low as 0.3 percent.

These faux vegetarians may be misleading themselves and others as a result of the same force that pushes meat eaters to argue with them: First, they claim to be vegetarian to reduce their own cognitive dissonance. This particular technique involves convincing themselves that they really *do* avoid meat, evidence to the contrary, since it makes them think they did change their behavior to match their do-no-harm values. Second, despite their best intentions, the meat cravings could just be too strong to resist. That's particularly true of health vegetarians. When people go "veg" for ethical reasons, they often become disgusted by meat, since disgust is a reaction to things we find morally offensive.

That doesn't mean that ethical vegetarians don't get tempted by meat. Some of them do. Take Richard Landau, Jacoby's famous chef husband. I meet him in Vedge's surprisingly small kitchen, where a few young cooks are busy peeling rutabagas, the earthy scent of the vegetables filling the air. Just like his wife, Landau defies the weakling-vegetarian stereotype. He talks and moves fast and seems to fill up the space with his presence. When I ask him about meat, he admits: "I miss it all the time," and then adds: "It's this meat craving that keeps me on my toes and makes me creative." Landau tries to recall the flavors of meat using vegetables and, as he calls it, "appeal to our campfire side." The way to do it is through smoking, he tells me. First, he tried using hickory smoke on his veggies, but that was "too one dimensional," so he moved to mesquite and apple chips. He also uses marinades. The closest to steak he managed to get so far was grilled seitan that he had soaked in a mixture of rosemary, balsamic vinegar, and peppercorns. "When you cook it, the rosemary gets all perfumy, and the balsamic vinegar caramelizes on the seitan. That really brings out the meatiness," he says.

Even though Landau keeps craving animal protein, psychology is on his side. It's easier to avoid eating meat, no matter the temptations, if the thing actually disgusts you, as it often does ethical vegetarians. And yet ethical plant eaters, Landau included, frequently fill their shopping carts and their plates with so-called fake meats: tofurkies, meatless meatballs, and veggie burgers. Is that another proof that even the strictest vegetarians can't get unhooked from the taste of meat? The answer is yes—and no. Yes, because there is indeed something unusually appealing in meat's potent mix of umami, fat, and the flavors of the Maillard reaction. That's what producers of mock meats are working hard to copy. And no, because quite often we keep eating meat simply out of habit.

Habit is a powerful thing. Approximately 45 percent of what we do each day is a habit—an action repeated in more or less the same way in the same place. If we were to make conscious decisions about all everyday behaviors, our prefrontal cortexes would screech under the strain. That's why we like habits: they are easier on our brains and on our nerves. Same goes for food habits, or what psychologists call "eating scripts." If we see a barbecue, we think burgers. If we are at a baseball game, we think hot dogs. And when we open our Sunday morning newspaper (or a phone news app), we think fried bacon. We like our habits, and we like the foods we already know. In experiments, people prefer tastes and smells they are familiar with over anything new. To boot, our meat habits get perpetuated by our surroundings, even, for instance, by "regular" families on TV eating their roasts, their burgers, their bacon. We simply follow their example.

As I stroll with Evelyn Kimber around the meat-smelling grounds of the Boston Vegetarian Food Festival, she tells me that it's actually the omnivores that are the prime audience for all these faux meats and veggie burgers. Such products can help them cut down on meat. "Many people wonder: If I weren't eating meat, what *would* I eat?" she says. It's easier to replace a beef patty on a barbecue with a veggie patty than to make something barbecue-able out of fresh veggies yourself. What's more, in Western cultures, we are used to a specific plate composition. Basically, there should be meat, a starch, and two vegetables on it. Take out the meat, and not enough is left. What should you cook instead? A dal? A vegetable tagine? It's simpler to just place meatless meatballs where regular meatballs used to be. And once you start using veggie beef or chicken, they become a habit themselves. So you buy more, no matter whether you crave the taste of meat or not at all.

Lack of cooking skills is also one of the main reasons why some vegetarians go back to eating meat and become—as the media call them—"born-again carnivores." In surveys they speak of plant-based cooking as "inconvenient" and "too much work." Chef Richard Landau agrees: "It's a lot of chopping," he says with a grin, then adds, "What's more, each vegetable has a specific level of doneness. Things like turnips—if they are underdone, they are stringy and bitter, and if they are overcooked, they get mushy." Does that mean you have to be a pro chef to satisfy your taste buds without meat? Of course not. But it does require some learning, and patience. Says Landau: "You have to give vegetables the same attention you would give meat. Really watch them cook, so they get perfect."

Another major reason why vegetarians turn back into carnivores is lack of social support. A typical ex-vegetarian is a young woman who has just moved in with her meat-eating boyfriend. Not only is cooking two meals too much hassle, but she also feels alone in her choice of diet. It's difficult to be the odd one, the one who has to explain all the time, argue, and convince. The need to belong—to feel that you are just like others and to share food with them—is a powerful incentive to keep eating meat and a difficult challenge to overcome.

And so it appears that some people in some situations may be more likely than others to give up meat. If your parents don't like animal protein that much, you are not neophobic, and you don't have the TT allele of the serotonin receptor genes; if you are liberal, open to experiences, and nonauthoritarian; if you are either single or in a relationship with a vegetarian; and if you find yourself in the midst of big life changes (moving to a new city, getting divorced), there is a higher chance you will cut out meat from your diet. It also appears that adding an ethical edge can make it easier to stop eating animal protein.

Although there is some indication that genetics plays a part in our choice of diet, scientists say that such DNA-based differences in food preferences are not big. What matters far more is the culture in which we grow up. It's difficult to give up meat because eating it is a habit, because we lack knowledge and skills for cooking vegetarian, and because pervasive vegetarian clichés don't encourage us to join the ranks of "soyheads." Furthermore, we have dissonance-reducing mechanisms that help us continue eating meat. We may think of animals as dumb creatures that cannot feel pain. We may cognitively erase the connection between a living animal and its flesh on our plate and rely on the language to help us do this. And, ironically, the more we doubt that meat eating is the right way to go, the stronger we may react to the vegetarians around us. We may even make them feel so uncomfortable—socially excluded, hassled, and tired—that they finally give up and sink their forks into a piece of beef or pork, or whatever kind of meat their culture pushes them to consume. Even if that meat may be dog.

10

Dog Skewers, Beef Burgers, and Other Weird Meats

Coconut-Cream Marinated Dog on Skewers (*Saté bumbu dendeng*)
This is a traditional Indonesian recipe.
Cut dog meat into pieces and marinate them in a mixture of coconut cream with a little soy sauce, pounded garlic and onions, ground coriander, ground cumin, salt, and pepper. Skewer, broil over charcoal, and serve with a pickled hot pepper sauce.

The Ituri Forest is an expanse of green springing from the fertile soils of the eastern Democratic Republic of Congo. It's a wild place, largely unexplored. The mysterious zebra-like okapi live here, as do groups of Mbuti and WaLese pygmies, hunter-gatherers of short stature (the average height of an adult is five feet, two inches). It was in the Ituri Forest, in 1981, that Richard Wrangham, the Harvard anthropologist who maintains that cooking made us human, got a lesson in the power of meat taboos.

Back in 1981, the Ituri Forest was not an easy place to live for a Western scientist. Wrangham, primatologist Elizabeth Ross (his wife), and their two colleagues sustained themselves for months on nothing but beans and rice. So when one day a Greek hunter passed by their camp and offered them two great blue turacos that he had shot, the scientists welcomed the chance to consume the kind of protein they'd been craving. But the WaLese people, among whom they lived, disagreed. Do not eat the turacos, they warned. It's dangerous. It's taboo. To the Westerners, the turaco was food, like chicken or turkey. And to their meat-starved taste buds, it was delicious. The next day, though, the two Western men developed diarrhea. "We truly suffered," Wrangham recalls. When the WaLese blamed the disease on the broken taboo, Wrangham pointed to the fact that not everyone who ate the turacos got sick. After all, the women were fine. "Oh yes, of course," the WaLese replied. "The taboo only applies to men."

There is nothing in the meat of turacos, at least as far as we know, that can specifically hurt male humans. Besides, other East African tribes have similar taboos on chicken. Yet no matter how much Wrangham tried to argue with the WaLese that turaco meat is fine for men, the Africans weren't convinced. That's the thing about meat taboos: although evident to those who practice them, they appear irrational to people from other cultures.

Africa was also where I received my first and most disturbing lesson in different cultural approaches toward meat. While Wrangham and his colleagues were the ones bewildered by a taboo, and the ones who broke it, I found myself on the other side of the taboo divide. Like the WaLese, I felt how deeply distressing it is to have your meat beliefs challenged and shaken.

It happened over twelve years ago in Limbe, Cameroon, on a hot, sticky night. Wads of mist were rolling into town from the massive slopes of a nearby volcano, Mount Cameroon, and the humidity seemed to intensify all the smells around: wet jungle, dusty roads, and smoke, which wafted from the grills that popped up the moment the sun set. I was hungry—and the meat skewers prepared by the locals seemed delicious (I was an avid meat eater back then). I approached one of the grills and asked the vendor what was on offer. Chicken, he said. Beef. Some soy. I raised my brows in surprise. Soy? Here, in a small Cameroonian town? Curious, I ordered one skewer of the grilled soy. But the moment I took a bite, I knew it was not what I understood to be soy. It had bones. It was meat. I asked the vendor for an explanation. He didn't understand what the problem was. After all, everyone knows that in the local language *soya* means "roasted meat," and if it's not chicken, goat, or cow, there is only one other thing it could be. *Rat.* I had just eaten grilled vermin! With some sauce and spices, but still. I was disgusted. If only I'd known, I would have never let this thing touch my lips.

If you are like me (and like most Westerners), you don't eat rats. You also don't eat horses, dogs, humans, or Egyptian fruit bats. You might, though, enjoy cows in the form of a steak and pigs in the form of bacon. You also probably wouldn't mind turacos, in a pinch. Yet, there are people around the world who would find some of these dietary choices weird or even disgusting. And it's not just the Hindus with their beef taboo or the pork-avoiding Jews and Muslims. To some Somali tribes, the idea of eating fish is revolting, while the Chuka of Mount Kenya would never dine on chicken and believe that if they did, they would turn as bald and pink as Europeans. Meanwhile in Asia, between thirteen and sixteen million dogs are cooked and consumed each year. Dogs, according to 83 percent of South Koreans, are meat.

The thing about meat taboos is that they tend to stir strong emotions. If you invite a dog-eating Korean and a dog-loving American to the dinner table and ask them to talk about their food preferences, the conversation will most likely be at least as heated as the one between a vegan and a carnivore. The question, it appears, is not just why we eat meat. It's also why we eat certain *types* of meat and hate or dismiss others. Why is beef in America, dog meat in Korea, and horsemeat in Kazakhstan craved and eaten while these same meats arouse disgust or are taboo in other countries?

As I discovered in Cameroon, and as Richard Wrangham discovered in the Ituri Forest, determining which animals are OK to grill and which should never enter our stomachs is far from fixed or universal. What is universal is that almost all cultures have food taboos, and no food is more widely affected by prohibitions than meat. That, too, only serves to highlight how important animal flesh is to us.

However, meat taboos can change over time. The story of horsemeat is a perfect example and proof that we can get hooked on new meats and unhooked off them again. When early in 2013 some Europeans discovered that beef products sold in their supermarkets were, in fact, made of horses, many were utterly disgusted. The outcry of the Irish was among the loudest. Professor Alan Reilly, the chief executive of the Republic of Ireland's food safety authority, asserted: "In Ireland, it is not in our culture to eat horsemeat." Yet it hasn't always been that way—the Iron Age Irish didn't see anything wrong in roasting a cut of stallion. They left archaeologists evidence of their dietary habits under the lush hills of central Ireland: plentiful horse bones, indicating that horses had been butchered and cooked.

But it was not just the Irish who over the ages changed their position on eating horses. In prehistoric times, much of Europe happily cooked its way through the stables—until the Christian church put an end to it. According to the Bible, hippophagy (which is the scientific term for eating horses) is a big no-no: "Of their flesh shall ye not eat, and their carcase shall ye not touch," says the book of Leviticus. To make things worse, the heathens sacrificed horses to their gods: the Angles of England did it, the Slavs did it, the Germanic Teutons did it. When the Christians set out to convert the barbarians, they decided that the horse sacrifices and horsemeat feasts had to go. Some of the pagans resisted. The Icelanders in particular were so committed to their horseflesh that the issue became a significant obstacle on Iceland's way to Christianity. In the end, rather than lose Icelanders, the church granted them an exemption. They still enjoy equine meat up there, especially in fondues. It gives the dish a desirable strong flavor, the Icelanders argue.

The rest of Europe gave up hippophagy with less resistance, and over the Middle Ages, eating horses was only done during times of famine. Even faced with lengthy sieges and hunger, people would rather eat grass and their own leather garments than horses. Those who tried the forbidden flesh in times of plenty were often severely punished. A ninth-century Irish handbook for confessors required that the eaters of horsemeat do penance for as long as three and a half years—longer than was prescribed for women who indulged in lesbian sex. Obviously, the roast-stallion feasts of the past were by then well forgotten, pushed into oblivion by Catholic guilt.

However, many modern European nations—the French, the Germans, and the Italians, for instance—feel little remorse over eating horsemeat. Attitudes changed in the nineteenth century. After the Industrial Revolution, the population of Europe doubled, and the price of meat shot up. While people were going hungry, thousands of horses were literally worked to death, collapsing from exertion while drawing carriages and trams, powering factories, and hauling coal from mines. Their carcasses would then be turned into glue, leather, and pet food. Across the continent, including Great Britain and France, the elites came up with a solution: let us convince people to eat horsemeat.

The arguments for consuming equine flesh were numerous: it's cheap, tasty, and nutritious. Not eating it, the reasoning went, amounts to a horrendous waste. It's more humanitarian to kill the old horses than drive them until they collapse. In the medical press, physicians recommended raw horsemeat and horse blood as a remedy for tuberculosis. Due to its relatively high levels of iron, the meat of horses was said to be of particular benefit to laborers, the anemic, and convalescents. Articles and books arguing the case piled up. In the end, the French *hippophagistes* succeeded in convincing their fellow countrymen to eat horses. The British failed. The reason? According to historians, several factors conspired to keep the British off horsemeat: lack of support from British butchers and restaurant owners; better access to international beef markets, lessening the need to eat horsemeat; and, unlike France, lack of support from the scientific elites in Britain, who didn't join forces to lead a horsemeat-eating movement to the same extent.

One of the big fiascos turned out to be a posh banquet at London's Langham Hotel in 1868, which was held to convince the British intelligentsia to eat horsemeat. The venue was magnificent: a truly palatial hotel that, according to contemporaries, had "no superior in Europe or America." A table for 150 guests was set under the vaulted ceiling in the Salle à Manger, the most exquisite dining room of the hotel. Champagne was plentiful, the waiters attentive and discreet. The menu promised ten dishes. To start, *le consommé du cheval*—a horse soup, followed by boiled withers. Once the guests—MPs, leading journalists, writers, and scientists—were seated, the principal organizer, Algernon Sidney Bicknell, stood up to give a speech. He talked about the unnecessary waste of perfectly good meat that could feed the masses. Seventy-five thousand horses, he claimed, toiled to their deaths each year in London. Let us butcher them and sell them to the poor. The guests applauded and began to eat. Among the diners was Frank Buckland, a surgeon by training and an eccentric celebrity by practice. Buckland was known across England for his peculiar culinary tastes: calling himself a "zoöphagist," he ate his way through the entire animal kingdom. He dined on Japanese sea slugs, boiled elephant trunks, stewed moles, toasted mice, panthers—you name it. If it moved and was made of protein, he had it cooked and served. Among the people gathered in Langham Hotel, Buckland was the person whose verdict on horsemeat would matter the most. The organizers had high expectations. Earlier, a similar "banquet hippophagique" in Paris proved an enormous success. Unfortunately, not so in London. Buckland was not pleased. A few days after the Langham dinner, he stated in a review: "The meat is nasty. I confess that I suffered tortures over which I will draw a veil."

Soon, the Langham banquet became the subject of jokes. In 1879, the *British Medical Journal* suggested forgoing horsemeat "'till English cooks are more skilled in concealing their raw material." In just a few years, the failed horsemeat revolution in Britain was turned into a cultural crusade against the French. The French, the reasoning went, ate horses because their culinary tastes were mercurial and indecent. The Brits, by contrast, felt superior. Today, equine meat is rarely found in the United Kingdom, and both Americans and Canadians picked up on the anti-French, anti-horsemeat sentiment. As one historian said: "Not consuming horsemeat became a marker of Anglo-American dietary self-consciousness."

Americans are at least as squeamish about the flesh of horses as the British are. Although horsemeat was briefly sold in the United States during World War II, today it is difficult to find. That's hardly a surprise. After all, many Americans are descendants of the Irish, who were trained by the Catholic Church to find horsemeat disgusting, and of the British, whose apparently inept chefs helped root the horsemeat taboo even deeper.

Meanwhile, millions of people around the world feel no revulsion over eating horses. China is the most profligate, cooking almost half a million tons per year. Mexico comes in second, and Italy third. Denizens of the horse-riding nations of Central Asia—the Mongols, the Kazakhs, and the Kyrgyz—see equine flesh as the most prestigious of meats. According to the Kazakhs, it does not spoil as quickly as beef and will not give you a stomachache. It is also perfect for weaning babies.

It seems that whether you find eating horses disgusting or not has more to do with where and when you were born than with the qualities of the meat itself. And the case of dog meat is quite similar.

On Sulawesi, an Indonesian island shaped like a fat girl with a long ponytail, eating dogs is nothing unusual. With light or dark fur, grilled or cooked, dogs have been eaten here for centuries. Yet anthropologist Daniel Fessler of the University of California, Los Angeles, who conducted research on meat taboos in Indonesia, could not stomach trying the local dog stew. Fessler, a slim, bearded man who looks like someone who has spent a lot of time working outdoors in exotic places, calls himself "largely vegetarian." Although he has eaten many things in his life that are not considered food in the West, such as insects and songbirds, dogs are just out of the question for him. "I tactfully avoided situations in which I would be offered dog meat, as I have a high regard for dogs' cognitive abilities and think they ought not be killed for food," he told me.

Among Westerners, eating dogs tends to stir even more emotions than eating horses does. To most, just as to Fessler, dogs are pets—period. They are to be pampered, not cooked. However, humanity's relationship with dogs and their flesh is not exactly straightforward. According to a theory put forward by Australian scientists, one of the reasons behind the domestication of wolves might well have been their meat. By the Bronze Age, dog eating was widespread in Europe. The ancient Greeks believed that dog meat helped with intestinal problems and itching. The flesh of a puppy eaten with wine and myrrh was said to cure epilepsy. Even the early North American settlers ate dog, and not just because they had no other food to put on their plates. Today, about sixteen million dogs are consumed each year in Asia alone. As long as it is properly cooked, dog meat is not bad for humans—or at least not any worse than chicken or beef. Its taste is often described as buttery, complex, or gamey. It contains about as much protein as pork but less fat. South Koreans, the biggest fans of dog meat on the planet, believe that it is good for the yang or the male, hot component of human nature. As such, many claim that eating dog meat helps the eater endure the heat and humidity of South Korean summers—you "fight fire with fire," the local saying goes. That is why the bulk of dog meat is consumed in South Korea during the three days that are traditionally considered the hottest—known as *chobok, jungbok*, and *malbok*—which come in ten-day intervals. Also, because dog is yang, it's mostly Korean men who eat it—92 percent of them report having tasted the flesh of a dog, compared to 68 percent of women. Meanwhile, South Korean women, if they so wish, can enjoy the supposedly healing properties of dog fat in several cosmetic products that have recently been released on Asian markets, from dog-oil cream to dog-oil essence and dog-oil emulsion. South Koreans also eat dog in the form of dog-meat kimchi, dog-meat-flavored mayonnaise, and dog-meat candy. South Koreans, it seems, like the taste of dog.

Many Westerners find it easy to condemn Koreans, Sulawesi, Thai, or Chinese for eating dog flesh. In the 1980s, French actress Brigitte Bardot launched a campaign against dog eating in South Korea. It didn't work. One thing the campaign did change, though, was the name given to Korean dog soup. In the restaurants of Seoul it is now sometimes called "Bardot."

Anthony Podberscek, anthrozoologist at the University of Sydney, has never eaten dog and wouldn't want to try it. But he conducted extensive research on dog-meat taboos. "In South Korea, dog eating is considered a major part of the culture, just as kimchi," he told me. "Calls from the West to ban the practice are viewed as an attack on the South Korean national identity. The lack of consistency in the behavior of Westerners toward cats and dogs leads to annoyance among South Koreans when they are criticized for consuming these animals." If South Koreans were to pick which animals should not be eaten, only 24 percent would say dogs, and 33 percent would choose cows. However, you will not see South Korean media waging war against the West's addiction to beef.

However, the real question at hand is about dynamics far more universal than nations' clashing opinions about dogs: Why is it that cultures differ so widely in defining which flesh is OK to cook and which should never touch their lips? Why do Westerners find farming dogs for food revolting while eating pigs or cows is considered OK? Why are they as eager to condemn Koreans or Sulawesi for eating dogs as vegetarians are eager to condemn all carnivores?

Most societies do not eat all the species local nature offers them, even if they are easy to hunt and their meat is nutritious. In Poland, the country of my childhood, eating *Helix pomatia,* or Burgundy snails, is a no-no. I remember those snails. They were almost everywhere: in our gardens, in the forests, on the sidewalks. They would have made an easy-to-get and cheap addition to the Polish cuisine, especially during the meat hunger of Communist times, when the shelves of Polish butcher stores were literally empty. But it didn't work out this way. Snails, according to the Poles, are slimy and disgusting. If the French want to eat escargots, good for them. And good for the Polish people too, since they can gather the snails and ship them to France in vast, profitable quantities—in total 230 tons per year. Bon appétit.

But it's not just the Poles and their abundant snails. The !Kung bushmen of Botswana consider only ten out of the fifty-four local wild animal species edible—even though, in theory, they all are. According to Fessler, who has studied meat taboos across seventy-eight cultures, Europeans hold most of them. North Americans are somewhere in the middle of the spectrum—not too squeamish but not particularly adventurous either. In general, Fessler believes, people don't give much thought to why they eat some species and religiously avoid others. "It's disgusting!" they'll usually just say, end of story. But to researchers like Fessler, Podberscek, and Wrangham, that isn't a good-enough explanation. The reaction of disgust is just the outer layer of a meat taboo, a cover. They have to peel it away and dig deeper.

As of today, many scientists still disagree on why we form meat taboos. The theories are many, and likely each contains a bit of truth. If you go up to someone on an American or a British street and ask them why we don't eat dogs, many will reply that it's because they are man's best friends. It seems plausible. Maybe sharing a couch with a dog or a cat makes us softhearted toward all the members of these species? Yet the case of the dog-eating, pet-keeping South Koreans proves this theory wrong. The majority of South Koreans don't think of dogs only as food. Almost 10 percent of them—at least in major cities—also keep dogs as pets. The worth of the rapidly growing pet market in South Korea has been recently estimated at US $1.3 billion. Surprisingly, pet owners are not much more likely to disapprove of using dogs for food than those who do not share their homes with animal companions—58 percent compared with 53 percent, respectively. South Koreans, it seems, compartmentalize some dogs as food, and some as man's best friends. Nureongi, midsized, yellow-furred dogs bred for their meat, belong in the first category, while Malteses, Shih Tzus, and Yorkshire terriers belong in the second. If you shop for dogs in South Korean markets, it is clear which canines are meant to be eaten and which to be cuddled. The clue is the color of the cage in which they are kept: pink is for pets, rust colored is for meat.

And it is not just the Koreans. The Oglala Sioux of South Dakota clearly separate which dogs are pets and which will be sacrificed to the gods and eaten. In Melanesia, pigs are treated like pets, sometimes even like human babies—women will breastfeed them, for example. And yet, they are still butchered and turned into pork.

If it's not the status of a pet that keeps the meat of some species taboo, maybe, as a popular claim goes, we just don't eat animals that are smarter than the rest—like dogs, cats, and horses. But that, too, does not seem to be the case. First, no matter how much their owners would like to believe it, cats and dogs are not exactly furry Einsteins. Although it's not easy to compare the intelligence of different animal species, we do know that pigs are at least as smart as dogs, if not smarter. It is not just the circus-type stunts that swine are good at—jumping hoops, bowing, spinning, and rolling out rugs. They can be taught to operate thermostats in their pens and adjust the temperature to their liking. They can press buttons and switch levers to get food. They can even master simple computer games. At Pennsylvania State University, two pigs, Hamlet and Omelette, have been taught how to operate a joystick with their snouts. Using M&Ms as rewards, scientists trained the pigs to move a cursor across a computer screen and line it up with other items. Hamlet and Omelette mastered that skill as fast as chimpanzees did in similar experiments. It appears that Sir Winston Churchill might have been onto something when he said: "Cats look down on you; dogs look up to you; but pigs look you in the eye as equals."

Cows may not qualify as computer-gaming pros, but they, too, can easily grasp how to operate the lever of a drinking fountain or press a button to get grain. Their social lives are surprisingly complex: they develop long-term friendships and may hold grudges against other cows. Even chickens are not exactly birdbrains. They make over thirty different types of sounds to communicate with one another. For example, they can inform the flock whether a predator is approaching by land or from the air. Plus, chickens know how to find and retrieve a hidden object—a test that quite a few dogs have failed.

We may believe that we choose to eat only the dumbest of animals, ones that cannot much comprehend what's happening to them anyway, but that is simply not true. If there was anything to it, then we should be making bacon out of dogs, not pigs. As Brock Bastian's experiments have shown, we may think swine and cows aren't bright precisely *because* we eat them—to quiet our cognitive dissonance over consuming animal flesh.

If it is not the cuteness, pet worthiness, or smarts of animals that makes their meat taboo, maybe over the centuries we have learned to avoid eating meat that is bad for our health. A popular theory states that people shun certain meats if consuming them could be dangerous. It is true that animals are breeding grounds for bacteria and parasites: roundworm, tapeworm, *Trichinella spiralis*, *Giardia duodenalis*, *Toxoplasma gondii*, *Escherichia coli*, *Salmonella enterica*—the list goes on. Eating dog meat may cause brucellosis and anthrax. Handling monkey meat can cause Ebola. Staying clear of dog and monkey is simply safer.

The Hebrew pork taboo is often explained as a means of avoiding trichinosis—a parasitic disease caused by the larvae of *Trichinella spiralis*. Once the larvae are ingested with undercooked meat, they can migrate in the human body. Fever, muscle weakness, or even stroke may follow. One problem with the health explanation for the Jewish and Islamic pork taboo is that trichinosis takes a long time to develop—too long for people without access to modern medicine to link the disease with its cause. It was only in 1859 that scientists made the connection between eating undercooked pork and trichinosis. And if the Jews and Muslims of the past didn't know pork causes trichinosis, why would they ban it? Besides, if eating pork was indeed so risky, why is it so widely consumed all over the world: in cold and hot climates, on the savannas, and in deserts and jungles? There is nothing extraordinary about the dangers of pork. Undercooked beef is dangerous too (it can contain tapeworm), and so is sheep meat, which can give you bacterial brucellosis or anthrax, a disease that often ends in death—unlike the usually mild trichinosis. Anthropologist Marvin Harris may have been right when he wrote: "If the taboo on pork was a divinely inspired health ordinance, it is the oldest recorded case of medical malpractice. A simple advisory against undercooking pork would have sufficed."

So why do we eat some meats and not others? Marvin Harris claimed that it all boils down to economics: meat taboos improve resource availability and help societies survive. That's why Hindus do not eat cows, and Jews and Muslims eschew pork. If the Jews and Muslims did keep pigs to eat them, Harris argued, the animals would compete with humans for grain and water—resources that are in short supply in the Middle East. At the beginning of the Neolithic period, swine were better suited for the climate of the Arabian Peninsula. Back then, the region was covered in dense oak and beech forests, which provided pigs with mud to wallow in and acorns and beechnuts to eat. However, when the population increased, the forests were chopped down. The shade, mud, and acorns soon became a distant memory. To keep pigs, you had to feed them grain and provide them with a lot of water to cool down their bodies. Since cows, sheep, and goats can thrive in a hot climate on little more than straw and bushes—things humans do not eat anyway—they proved a better choice as livestock. Pigs not only competed with people for the same resources but also could not be milked. They became costly, so they had to go.

A similar theory helps explain why cows are sacred to Hindus—butchering them would be just uneconomical. Four thousand years ago Hindus not only killed cows but also ate them. The earliest Vedas (holy texts) did not forbid the slaughter of cattle; only around 1000 CE did cows become sacred, similar to the way they are now. Today, India is overrun with cows. There are cows in the markets, wandering between the stalls, sleeping across the tracks on railway stations, grazing on trash in front of restaurants. I once saw a cow lounging on the steps of the swanky offices of Ernst & Young in New Delhi. What changed over the centuries so that cows ceased being meat and achieved a holy status? According to Harris, the answer is again rooted in economics. It started with a population explosion, which precipitated the clearing of woods for fields. As the previously forested Ganges Valley turned into barren land, drought became common and agriculture difficult. "The farmers who decided not to eat their cows, who saved them for procreation to produce oxen, were the ones who survived the natural disasters," explains Harris. Oxen pull the plow, cows give milk, and both produce dung, which fertilizes the fields of India and fires its stoves. It's been calculated that in modern India, the dung used as fuel for cooking is equivalent to forty-three million tons of coal (that's much more than Canada exports each year). And so, as Harris writes: "Those who ate beef lost the tools with which to farm. Over a period of centuries, more and more farmers probably avoided beef until an unwritten taboo came into existence." Similar reasons strengthened the church-imposed horsemeat taboo in medieval Europe. Horses, as opposed to chickens, are not good at converting food into muscle. They are simply not efficient meat machines, but they are quite useful when alive: for transportation and for plowing fields and fertilizing them with manure.

Although the sustainability theory solves some of the riddles of meat taboos, it doesn't solve them all. Can it be that we do not eat dogs in the West just because it's uneconomical? Not really. After all, eating all the strays would make perfect economical sense. And what about the aversion of some Somali tribes toward consuming fish, a taboo that can't be explained with economical or environmental reasons? Some such tribes live by lakes or rivers brimming with fish, and yet they fail to profit from this protein abundance, believing that consuming fish would cause their teeth to fall out. And what about the turaco taboo among the pygmies of the Ituri Forest? Wrangham believes that such taboos make people feel they belong. When he worked in East Africa, he tells me, he encountered food taboos at many levels of social organization: "They had different taboos for different subclans, and for different clans, and for different tribes. These taboos acted as markers of identity, and that is why they often applied to men and not to women, because the clans were based on male kinship, whereas women moved around." Other scientists agree with Wrangham and emphasize the role of taboos as a marker of cultural distinction: observing meat taboos can help people feel that they are part of a group. If you don't eat dogs, you can shake your head with other fellow Americans at those brutal Sulawesi and South Koreans. If you don't eat turacos, you are tied with other WaLese men. One of the explanations behind India's cow taboo is the rise of Islam and the need of Hindus to separate themselves from the Muslims. In a similar pattern, the pork taboo in the Middle East helped differentiate Muslims and Jews from the Christians. And in more recent times, the horsemeat taboo gives the Britons and Americans another weapon against those weird French. By forbidding a food that is so highly nutritious and desirable as meat, a group of people can set themselves apart from their neighbors and feel united. If you were starting a new religion, for example, prohibiting a popular meat could give you an edge, something like a brand. In a similar way, not eating any meat whatsoever can also help people experience the pleasure of belonging: they are a tribe of vegetarians, set apart from the tribe of turkey-eating, burger-grilling carnivores. It goes the other way, too, of course. If you are committed to eating animals, you can roll your eyes with your buddies (over a steak or a turaco wing) at all those silly tree-hugging vegetarians. You know which tribe you belong to.

For the time being, almost all human cultures embrace consuming at least a few species of animals. Some—like Asians—eat more types of meats; others—like Americans and Europeans—a few less. But cultures change. They evolve. As the example of meat taboos shows, our meat-eating habits adjust to the realities of our economies and the state of the environment in which we live. For the Hindus of India, eating cows became too damaging, so they made it taboo. For the Jews of the Middle East, raising pigs became contrary to their best interests—and so they banned it. That's not the whole story, of course, but an important part of it nevertheless. As our planet's climate undergoes fast and negative changes, will our meat taboos evolve, too? Will we start eating insects and stop eating cows? Or will we, like the vegetarians of today, make all meats taboo? For now, in some parts of the world at least, the trend is disturbingly reversed: longtime vegetarians are getting hooked on meat again, giving up plant-based diets and consuming more and more flesh, no matter the health and environmental consequences.

11

The Pink Revolution, or How Asia Is Getting Hooked on Meat, Fast

As steak houses go, The Only Place is rather unassuming. The atmosphere is relaxed, the decor quite simple: square tables covered in red-and-white-checkered linen, alpine-style wooden chairs. Rows of Christmas lights blink lazily under the ceiling. The mustachioed waiters appear slightly bored in their crisp white shirts. In fact, little would be notable about The Only Place if not for its location. The steak house is hidden on one of the backstreets of downtown Bengaluru, India—the country where cows are so worshipped that killing one can get you in prison. In fact, their urine is considered sacred and is used to bathe sick kids. And yet the menu at The Only Place features Philly cheese steak, chateaubriand supreme, and double filet mignon. Beef, beef, and more beef.

My steak arrives on a simple white plate, in a cloud of succulent aromas. Although I take only a few bites (my husband polishes off the rest), it's enough to tell the meat is delicious. Sacred or not, Indian beef tastes good.

Whenever I tell this story, people (Westerners) usually react with shock: You had beef in India? Is it even legal? The answers are "yes" and "yes." The Only Place was a pioneer when it swung its doors open in 1970, but nowadays steak houses are all the rage in wealthy, metropolitan India. What's more, India has recently taken over Australia as the world's second-largest exporter of beef—after Brazil. That's right: the India of holy cows exports more dead bovines than almost any other nation on the planet.

Granted, India still has a very low meat intake—just 7 pounds per person per year to the US's astounding 275 pounds—but it's growing shockingly fast. By 2030, for example, poultry consumption in India's sprawling cities is projected to shoot up from 2000 levels by 1,277 percent. And the same thing is happening all across Asia. By 2030, Malaysia's beef consumption will likely be up 159 percent, Cambodia's 146 percent, and the urban dwellers in Laos will go through 1,049 percent more poultry. Meanwhile, China will gobble 22,050,000 tons more pork—that's about as much in weight as a hundred thousand fully loaded Boeing 787-8 Dreamliners. With 2.5 billion people in India and China alone, Asia's growing appetite for meat spells trouble not only for the animals that will be killed and eaten but also for the health of Asians—and of our planet.

What is happening in the nations of Asia and in developing countries on other continents, too, even if on a smaller scale, is what scientists refer to as the nutrition transition. There are five recognized patterns or stages of nutrition transition. First, a society goes from stage 1, collection of food (hunting and gathering), to stage 2, famine (which starts with agriculture). Then comes stage 3, receding famine, during which agriculture improves and severe hunger becomes a thing of the past but foods remain unprocessed and simple. Later, as time passes, societies go through an industrial revolution and enter stage 4, degenerative disease. That's where the West is now: eating poor diets loaded with cholesterol, sugar, and fat. But that's not the end of the path. Nutritionists predict that there is one more step to take: stage 5, behavioral change. Behavioral change means, in a way, moving back to eating foods similar to those consumed by stage 1 societies: much less meat, more fruits, veggies, and whole grains.

For now, what's going on in Asia is a transition from stage 3 (receding famine) to stage 4 (degenerative disease). The more money people in developing countries have to spend on food, the more meat they buy. One study showed that each increase in yearly income of US $1,000 boosts per capita meat consumption in Asian countries by 2.6 pounds, in Africa by 3.6 pounds, and in the Middle East by 8.8 pounds. Members of developing nations, which formerly comprised the world's least meathooked populations, have revealed that they're willing to spend their hard-earned wages on meat, which they may have no tradition of eating, and damage their health in the process. We're left with one question: Why?

The first East Asian country to develop an appetite for meat, and one that can offer a glimpse into the process of going from almost vegetarian to meat loving in a relatively short period of time, is Japan. As late as 1939 a typical Japanese ate just 0.1 ounce of meat per day. That's a yearly average, of course. Today, the daily meat portion of a typical Yamada Tarō (the Japanese equivalent of John Smith) is 4.7 ounces, and his favorite animal protein is pork, not tuna in a sushi roll. One reason behind this astounding change was the rise of Western influence.

Medieval Japan was practically vegetarian. The national religions, Buddhism and Shintoism, both promoted plant-based eating, but what was likely more key to keeping the Japanese off meat was the shortage of arable land on the islands. Growing livestock takes land away from more efficient plant agriculture, and already in medieval Japan, too many forests had been cleared for fields and too many draft animals were being killed for their flesh—which prompted Japan's rulers to issue meat-eating bans. The first such ban was announced in 675 CE and meant no beef, monkey, chicken, or dog in Japanese pots from late spring until early autumn. Later, more bans followed. For some time, the Japanese could still satisfy their meat cravings with wild game, but as the population increased and forests gave way to cropland, deer and boars disappeared and so did meat from the plates of the Yamada Tarōs.

The winds of change started blowing, at first mildly, in the eighteenth century. It was the Dutch who sowed in Japanese minds the idea that eating meat is good for health. The Japanese came to see the meat-loaded diets of the tall Europeans as a symbol of progress, of breaking with feudal, hierarchical society. In 1872, Japanese diets took a fast swerve toward meat. That year, on January 24, a feminine-looking, poetry-writing emperor Meiji publicly ate meat for the first time, giving the nation permission to follow his example. Over just five years, beef consumption in Tokyo shot up more than thirteen times (what made it possible were imports from Korea). Meiji and his government saw meat not only as a way to modernize Japan and boost the health of the average citizen but also as a way to bolster the strength of the Japanese army. Back then, typical conscripts were small and thin—over 16 percent of candidates failed to meet the minimum height of four feet eleven inches.

The American occupation after the Second World War gave another powerful boost to the Japanese hunger for meat. The Japanese observed the war victors stuffing themselves with hamburgers, steaks, and bacon. The words of Den Fujita, the chief of McDonald's Japanese operations, sum up the prevailing sentiment pretty well: "If we eat hamburgers for a thousand years, we will become blond. And when we become blond we can conquer the world."

The story of India's newfound taste for animal protein is in many ways similar to what had happened in Japan. It's a story of longing to become modern and powerful, a member of the "we've made it" club. And Bengaluru, this dusty tangle of humans and buildings and cars, this cacophony of a city, where the twenty-first century mixes with the distant past on every corner, is a perfect place to study India's ambivalent relationship with animal flesh.

Bengaluru used to be called the "garden city." But in recent years the trees and lawns have given way to offices and apartment buildings, to an outbreak of stores and potholed streets. Now Bengaluru is a city known for its pollution and congestion—but also for being the capital of the nation's IT industry, the silicon valley of India. It's loud. It's overwhelming. Expensive limousines maneuver between dirty, overcrowded buses and rusty auto rickshaws.

Renovated fancy boutiques at street level are housed in buildings that are otherwise crumbling. Loose wires hang over outdoor café tables, where middle-class patrons sip lattes. The air smells of perfumed women, of sweat, of gasoline and dust, the same dust that keeps pushing its way into my mouth and eyes.

The Only Place is much more peaceful. Behind its doors it is quiet enough so that I don't have to struggle to hear the story that Ajath

Anjanappa tells me over a beefsteak—the story of how so many young, middle-class Indians go about giving up vegetarianism.

Anjanappa is, in many ways, a typical successful Bengalurean. A thirty-something engineer with an MBA degree earned in the US, he now runs his own company, which provides energy-efficient lighting to local industries. He is easygoing and good-looking. And like many of his generation, he loves meat. These days, as Anjanappa explains, to eat steaks and burgers in India is to be modern and worldly. It is a sign that you belong to the group of people who jet around the planet and work for multinational corporations. "It can help you in your career," he tells me.

Anjanappa's affair with meat began the way it usually does for wealthy Indians: you grow up either in a family that is pure veg (Indian for "vegetarian") or one that consumes very little meat. You go to college, you make new friends. You start eating out in the many international restaurants that have sprouted all over Bengaluru, Mumbai, and Delhi. "All my non-veg friends were pushing me, saying I was missing out. If you are a vegetarian you don't belong to the same social circle. By the time we graduated, all my friends were meat eaters," Anjanappa says. Then you go to work for a Western corporation. "A lot of companies like Google or Apple have their own cafeterias where there is a lot of meat served, for free. So why not eat it?" Anjanappa tells me. As years pass, you travel for work outside India where there is often no decent vegetarian food to be had. You either eat meat, or you go hungry. So you eat it, and you start to like it. Those young Indians who work for multinational companies often make good money, and they spend it trying new things, including new cuisines. What makes pressure to eat meat harder to resist in India is the communal way in which meals are enjoyed there. As in many Asian countries, and not in the individualistic West, dishes in India are shared. There is a huge pressure to eat what others are having. To refuse food is to be antisocial.

Yet even in up-and-coming Bengaluru, steak houses are not a common sight. And many locals, when I ask them about their country's beef industry, are surprised to learn that India is the second-largest exporter of meat in the world. Unbeknownst to many, in 2013 and 2014, beef shipments from the subcontinent rose a staggering 31 percent. A big chunk of that meat goes first to Vietnam and later on to China. A lot ends up in the Gulf states and in North Africa. For the US and UK, there is only a trickle left, so the chances you are grilling Indian bovines on your barbecue are slim. But many Western producers are nevertheless worried about the competition. Indian beef is cheap and lean, and it's flooding the markets.

But at least officially, India is not killing off its holy cows. The beef it exports actually comes from water buffaloes, a species of bovine closely related to cows but not quite the same thing. Water buffaloes are not sacred. They do not spend their old age in senior houses or get buried in cemeteries—the way cows do in India. Instead, water buffaloes are overloaded on trucks, transported without food and water, and slaughtered in miserable conditions similar to the American meat industrial complex.

Officially, cow meat is not exported from India, but in reality, there exists an underground cow slaughter industry that labels the meat "buffalo" while it is still in India and relabels it "cow" once it crosses the border. According to a local chapter of the nonprofit organization PETA, there are about thirty thousand illegal slaughterhouses in India, many of them turning holy cows into steaks. As Anjanappa tells me: "As long as the money is coming in, they don't mind what they are killing."

For Western ethical vegetarians, who gave up meat for the sake of animals, such duality is often hard to comprehend. How is it possible that killing a holy cow is a horrible sin and often a crime, while butchering their close cousins, the buffaloes, is perfectly fine? How can young, wealthy Indians chuck their vegetarianism so fast—and seemingly with little regret? And yet, it does make sense. After all, in India, vegetarianism means a very different thing than it does in the West.

First and foremost, vegetarianism and the sacredness of cows don't necessarily go hand in hand. Millions of Indians who never touch beef have no problem at all eating chicken or pork. Meanwhile, even though cows have been considered holy in India for centuries, before 1000 CE the sacred animals were still slaughtered and eaten. The ban on eating holy cows crept into the culture slowly, over time. Today, devout Hindus believe that the body of every cow is inhabited by 330 million gods, and to become a cow, a soul has to transmigrate eighty-six times (that's a lot of lives to go through). Until recently, killing a cow carried the death penalty in the state of Kashmir. The fact that the ancestors of Indians ate the sacred animals is something that many in India are trying to forget: in 2006, mentions of ancient Hindus consuming beef were deleted from school textbooks.

Vegetarianism in India didn't arise from the veneration of cows. It developed independently, from a concept called *ahimsa*, or nonviolence. Basically, *ahimsa* means that all life is sacred and should not be destroyed. Nonviolence is a common thread linking the three Indian religions: Hinduism, Jainism, and Buddhism. But *ahimsa* doesn't necessarily imply that you cannot eat meat—not according to everyone's interpretation. *Ahimsa* is not about animals. It's about people. Just like Pythagoras in ancient Greece, Buddhism, Hinduism, and Jainism concentrate on what violence does to human souls and how it can degrade them. So if you didn't kill the animal yourself and didn't ask anyone to do it, there is no violence to stain your spirit, and you are OK. Buddha ate meat, and even the founder of Jainism, Mahavira, once ate a few pigeons that were killed by a cat.

For many in India *ahimsa* still means that meat should never be eaten because it comes from soul-polluting violence. Yet this form of vegetarianism is still all about humans and not about animals, which makes it easier for adherents to start eating meat once they cease being religious. There is very little discussion about the suffering of meat animals in Indian media, barely any at all. For most, vegetarianism is a way of life, a tradition taken as is. Those who consciously decide not to eat animals either for health or for ethical reasons stand out—so much so that they are called "out-of-choice vegetarians."

If vegetarianism in India is not a choice for most and often seen as conservative and even backward, meat eating stands for modernization and progress. Just like the Japanese poet-emperor Meiji, Mahatma Gandhi also, at some point in his life, believed that a meat-based diet could push India forward and upward. Yes, that Gandhi. Violence-abhorring, meat-abstaining Gandhi. But he wasn't always a vegetarian. Although he was born into a typical Indian religious pure-veg family, he soon came to regard meat eating as something that could help modernize the subcontinent. It was a common belief back in the late nineteenth century that to eat meat was a patriotic duty, a way for Indians to become as strong and powerful as the English so that they could drive the colonialists out. One comic poem became particularly popular: "Behold the mighty Englishman / He rules the Indian small, / Because being a meat-eater He is five cubits tall." And so, one day, Gandhi decided to start eating meat. He and his friend, already a meat eater, packed freshly baked bread and some cooked goat and hiked to a lonely spot by the river so that no one would see them. Once there, Gandhi took a bite of the meat and started to chew. He didn't like it, not at all. The meat was tough as leather. He just couldn't finish it. Later, at night, he was tormented by nightmares and felt as if "a live goat were bleating" inside him. But he kept reminding himself that eating meat was a duty, that he simply *had* to do it.

Soon after, when he moved to London, Gandhi began to eat meat in fancy restaurants and learned to enjoy the taste. He admitted in his autobiography that for a while he wished that "every Indian should be a meat-eater." But in time, Gandhi went back to vegetarianism. He became, as he said, a "vegetarian by choice." After reading many books on nutrition and ethics, he was convinced not only that vegetarianism was better from a moral perspective but also that a plant-based diet was advantageous for health. He started to see that meat eating wouldn't turn India into a powerful nation—and maybe just the opposite.

Today, you won't hear many Indians say that meat eating is a national duty and that it will help India rule the world, but the belief that animal protein makes individual people strong is still alive and well. In India, the media are rather quiet about all the studies that connect a meat-eating diet with cancer, diabetes, and heart disease; yet there are many articles that glorify meat eating as a key to good nutrition. The protein myth is particularly potent. The *Times of India*, a leading newspaper, writes: "Keep in mind that vegetarianism comes with its share of problems because plant foods tend to lack protein." On television, celebrity chefs whip up meat-based dishes, and male actors go non-veg to gain muscles for their roles. Salman Khan, the highest-paid Bollywood star, famed and beloved not only for his stage talents but also for his muscular body, is a proud meat eater, preaching the gospel of non-veg to the masses. And so it's hard to blame Anjanappa for believing that as a regular gym-goer he needs to eat meat—even though science clearly shows that that is not true. In fact, the traditional veg Indian fare is far from deficient in proteins. Just think about it: they have over fifty varieties of lentils, peas, and beans, all loaded with protein, and if paired with rice, the combination makes a complete protein. Vegetarianism could have taken root in India much easier than in Europe or North America precisely because of the culinary diversity and protein load that local plants could offer. A prominent Indian food historian, K. T. Achaya, went so far as to declare: "Perhaps nowhere else in the world except in India would it have been possible 3000 years ago to be a strict vegetarian."

Just as in Gandhi's times, though, consuming meat in India is still often a political act. Several beef-eating festivals organized at India's universities ended in violent clashes with more conservative groups. "Beef is a symbol of anti-Brahmanism," stated one student organization. And scientists agree: eating beef in India stands for modernization. It opposes the caste system, which is topped by beef-avoiding Brahmins, and thereby undermines authority. During the campaign for parliament in 2014, the conservative BJP party tried to win votes for its leader using a slogan: "Vote for Modi, give life to the cow." Narendra Modi, a stout man with white hair that connects via sideburns with his trim beard and perfectly circles his round face, coined the term *pink revolution* to describe India's growing hunger for meat—and for money made from exporting beef. Before the elections, Modi spoke of the meat industry's crimes against "mother cow" and suggested the beef trade should be banned. Yet months after he won and became prime minister, the pink revolution was still rolling. It seems that meat exports are just too good of a cash cow for India to shut down.

Beef may be the most politically sensitive of India's meats, but it's chicken that is the most often eaten. Girinagar, an upscale neighborhood of Kerala's twin city Ernakulam-Kochi, is leafy and green, with traffic subdued, almost calm. The air here is thick with foul vapors that rise from the area's many canals, green streams of trash and vegetation that crisscross this part of urban Kerala, but most houses are neat, oversized, and obviously pricey. In the maze of unnamed, narrow streets, a small store is hidden. A store that embraces India's growing hunger for meat and what's behind it: hunger for modernity, for riches. The store belongs to a chain called Suguna Daily Fressh and offers hygienically packed, easy-to-prepare chicken. It may look like nothing special to a Westerner, but in India, a country where until recently to cook a chicken entailed buying it at an overcrowded market, plucking it, and cleaning its guts, the store is very different. Stores like Suguna Daily Fressh, with their conveniently packaged and expensive products, are for the upper classes, who are driving India's desire for meat. After all, people crave meat precisely because it is expensive. Who doesn't want to be like the beautiful, light-skinned, and obviously successful people who gorge themselves on animal protein in the advertisements for KFC, McDonald's, and the Meat Products of India? India wants to go up, and to go up means to eat meat.

Similar things are happening in China. In a country where the average person ate less than seven pounds of meat *per year* (in the early twentieth century), China is fast becoming a nation where many plates are overflowing with pork, chicken, and, to a lesser extent, beef. Since the 1980s, meat consumption in the People's Republic has quadrupled. China is already ingesting over half of Earth's pork, 20 percent of its chicken, and 10 percent of its beef. Soon, these numbers will be much higher. If China, with its population of 1.3 billion, ate as much animal flesh as Americans do today, they would be hogging (no pun intended) over 70 percent of the meat produced on the planet.

The themes behind China's growing appetite for meat are similar to those in India and Japan. The Chinese are eating more and more meat because they are finally starting to be able to afford it and because the many years in which there was not enough animal flesh to go around resulted in it symbolizing luxury and wealth, modernity, the West, and power. And as in India, to eat meat in China often means to reject old social hierarchies. This is one of the reasons why fast-food joints such as McDonald's and KFC thrive in the People's Republic. As anthropologist Yunxiang Yan once wrote: "Many people patronize McDonald's to experience a moment of equality." In Western fast-food chains, all customers are treated with similar respect, no matter their age, social status, or wealth. That's very different from traditional Chinese eateries, where—as Yan described—there is an ongoing competition among customers as to who will order the most expensive, luxurious meal. Say the guy sitting at the table next to you orders a chicken. Now, if you don't want to lose face, you can't just have veggies. To prove your social standing, you have to order meat, too, and preferably a more expensive dish. You are just about to ask the waiter for pork, when the guy at another table, to your left, beats you to it and places an order for pork dumplings. After mentally calculating how many yuans are left in your wallet, you order, with a slightly shaky voice, the most expensive pork on the menu. Face saved, money lost. At McDonald's or KFC, with their short, simple menus of similarly priced and standardized dishes, such dilemmas don't exist. What's more, in these restaurants, people are consuming not just food but also Western culture, so often associated with individualism and democracy. And, of course, McDonald's, KFC, Burger King, and the like are all about eating meat, so if you go there to experience the spirit of Western equality, as a side order you may get hooked on burgers and chicken.

Even though there may be many parallels between India's and China's growing desire for meat, there are quite a few important differences, too—differences that are the reason why China is already gorging on much more animal protein. Although the traditional Chinese diet is mostly plant based, vegetarianism never took root in China as deeply as it has in India. The Chinese ate so little meat in the past largely because they simply didn't have enough good land to grow feed for livestock. Famines were common, very common, so the people here learned to eat anything that was available (hence the donkey penises, grilled scorpions, and other unusual-for-Westerners foods on Chinese menus). In China, the seeds for its future love affair with meat were already present, buried deep in the culture, waiting for a good time to sprout.

For many centuries, most followers of Buddhism in China were not vegetarian. By the sixth century, though, meat eating evolved to become a no-no for devout Chinese Buddhists. Consuming meat, the holy scriptures said, inhibits the ability to feel compassion. It causes nightmares. And yet Buddhist vegetarianism in China never became as widely practiced as Hindu vegetarianism did in India. It was mainly the domain of monks, while the elites kept wolfing down meat—so much of it that they were dubbed "the meat eaters." The rich of China were at least as extravagant in their meat tastes as were their European counterparts and ate things like yak tails, bear paws, and leopard fetuses. And just like the medieval European peasants, who hungered for all the animal flesh heaped on the plates of the aristocracy, the Chinese masses dreamed of eating just like their nobles and equated wealth with meat. (By contrast, Indian nobility meant vegetarianism, and so that's what people aspired to.) What also played a role in keeping China carnivorous, despite the arrival of Buddhism, was its bureaucracy. To rise through the ranks of the imperial government, a civil servant basically *had to* eat meat. Vegetarianism was considered inappropriate for high officials because public occasions required sharing meat to ensure social harmony. Meat avoiders were looked upon with suspicion and sometimes even forced to eat pork to make certain they didn't belong to some radical vegetarian sect (heretics in Europe come to mind). And if that wasn't enough to ensure China didn't go completely veg, the beliefs of traditional Chinese medicine discouraged plant-based diets, too. Vegetables are generally "cold," according to traditional Chinese dietary therapy. That's fine if you are suffering from a fever, but if you have chills or fatigue, you need to nourish yourself with "hot" foods—such as meat. In traditional Chinese food therapy, meat is often necessary to balance your energy. If you don't eat it, you may end up in trouble.

And then there is the Mandarin language. Take the words *chicken* and *fish*, for example. These words, when spoken in Mandarin, have the same sound as the words for "prosperous" and "abundance." For this reason, people in China eat chicken and fish on Lunar New Year's Eve to ensure good luck. *Home* is another word that may help reinforce meat-based diets in China. To make the character for "home" you basically take the character for a pig and put the character for a roof over it: pig plus roof equals home.

I remember staring at the thicket of Chinese characters covering the menu in one of Beijing's sprawling restaurants. The room was bright and open and dotted with round tables. Everything seemed loud in there: the people, the smells of frying and roasting, the colors—a jumble of white, red, and gold. The plastic menu, sticky from too many hands touching it before me, was in Chinese only, and the pictures were quite blurred. When a waiter walked past me, I caught his attention. "What is that?" I asked, in my stiff, phrase-book Chinese. "Meat," I heard in reply. "What meat?" I pressed. "Meat," the waiter shrugged. I pointed to another dish, then another, and kept asking. But my understanding of the food on offer got only slightly better. It appeared that although some dishes were "chicken" or "fish" or even "donkey" (there was a donkey penis soup listed), lots of others were just "meat," period.

In time I came to learn that if something in China is described simply as "meat," it means pork. The Chinese love pork. Every other pig that is alive on this planet is being raised in China—and it will be slaughtered and cooked there. In the famine-ridden past, pigs were economic security. They were cheap to raise, feeding on household leftovers and even human excrement, and could be exchanged for political favors, given as wedding presents, and—of course—eaten. Today, China's swine are still seen as a measure of food security, even on a national scale. The Communist government makes sure people keep buying pig meat by handing out grants and subsidies to hog producers, waiving their taxes, and offering them insurance. For China's leaders, providing citizens with enough pork to fill their plates means progress and modernity. It means they've succeeded.

But there are problems. Asia's nutrition transition means the health of people there is going downhill. It's not all on account of meat, of course. The sodas, the sweets, the fries—all this has a role too, an ugly one. Still, there are plenty of studies that connect high intakes of meat with higher odds of cancer, diabetes, and heart disease. And that's what Asia is getting right now. There are already over sixty-one million people with type 2 diabetes in India, and by 2030 that number will likely double. In the twin city Ernakulam-Kochi, where hygienically wrapped Suguna Daily Fressh chicken is sold, almost one in five inhabitants is diabetic. The waists of Asia are expanding, too. Over 30 percent of Chinese adults are overweight. One hundred million are obese.

The poor nutrition and resulting poor health are only part of Asia's meat-related problems. In March 2013, the pale bodies of over sixteen thousand dead pigs floated down Huangpu River near Shanghai. The carcasses were swollen and rotting, the stench nauseating. The animals, which may have died either of a virus or of extreme cold, had been dumped into the water from industrial farms upstream. The "hogwash incident," as it was called in the media, is just one of many scandals that have plagued India's and China's booming meat business. There was the "instant chicken" scandal (China), when poultry were supposedly given eighteen different antibiotics to grow ultrafast. There was "Avatar meat" (China), when pork was said to be contaminated by phosphorescent bacteria and glowed blue in the dark. There has been one avian flu outbreak after another. It's the sheer scale of the industry's growth that causes these problems. When it comes to animal products, China in many respects has more stringent safety regulations than the US does. Take ractopamine, a drug that mimics stress hormones, which is given to as many as 80 percent of American pigs. But when the Chinese found out it was administered to their hogs, a scandal erupted (ractopamine is illegal as a feed additive in China).

Asia's burgeoning appetite for meat is not just its problem—it's our problem, too. The meat industry is international, and what happens in one part of the world often affects the others. The major challenge for the industry in China is land—as in, there's not enough of it. China has a mere 0.08 hectare of arable land per person—6.5 times less than the US, over 16 times less than Canada. The Chinese simply can't grow feed for all the animals they want to eat. India is also struggling with a land shortage and with a severe water shortage to boot. If China and India want to have a meat industry, which is enormously water intensive, on an American scale, they will be in trouble.

What do countries do that want meat but can't produce enough of it? They could issue meat-eating bans the way medieval Japan did, but that's highly unlikely these days, of course. Instead, they outsource. They import. And where are all these chops and burgers going to come from? The US, for one. In the last decade, the flow of pork from the States to China rose almost ten times. And that was before the biggest meat processor, Shuanghui International, purchased the American giant Smithfield Foods, to become the world's largest pork producer. Yes, the money is pouring into US coffers, but there is a dark side to the deal. While the Chinese are consuming the meat, we are consuming the pollution: the lagoons of manure, the dirty air, the antibiotic-resistant bacteria.

The Chinese also import vast quantities of feed for their domestic livestock—and again, export the pollution. Chinese meat producers are on the constant lookout for land to grow soy and corn to fill the stomachs of their livestock. A lot of it comes from the US, some comes from Africa, and some from Eastern Europe, but the majority comes from Latin America. Already over 80 percent of Brazil's soy exports are going to China, and the growth curve is nearly vertical. A slice of Brazil the size of Colorado is currently covered in soy crops destined for China; a similar thing is happening in Argentina. That's not exactly good news. Ninety-nine percent of the soybeans grown in Brazil are genetically modified, and they are intensely sprayed with herbicides and fungicides. In Argentina's soy-growing districts, such use of chemicals has already caused epidemic levels of cancers and birth defects.

It's easy to criticize Asia's meat hunger and point fingers at the trouble it's causing, but these nations are basically following the path Europe and North America took some time ago. They are getting hooked on meat for many of the same reasons we did: because of meat's taste, because of the meat industry's lobbying and marketing, and because of meat's symbolism. They want meat because they want to be modern, industrial, and rich. Often, they want to get rid of social hierarchies—and the West with its meat-laden cuisines stands for equality. The power of meat's symbolism is particularly clear in India: when the Brahmin elites sat on top of the Indian world, the masses aspired to be vegetarian. Now, there is the West to look up to as the ultimate "we've made it" people. And these "we've made it" people eat plenty of meat. The young upper classes, the IT workers, and those with MBAs from US schools don't want to be like the traditional villagers with their stomachs filled with lentils. The media is selling the protein myth, and they buy it. Even when they were vegetarian, they didn't care for animals much, and they didn't choose their diets themselves anyway, which now can make going non-veg a bit easier.

The Chinese, meanwhile, have loved meat, and pork in particular, all along—they just didn't have enough land and resources to grow it. Now they can virtually "import" land from Brazil or the US, and they do. The Communist Party is all for it. Meat means prosperity, and the party wants the Chinese people to feel prosperous. The government distrusts vegetarianism because it is tied with religious movements of the past—something the government has worked hard to suppress. Asia is starting to eat more meat often not in spite of its vegetarian religions but precisely because of them—as a way to reject them and consign them to the past.

Of course, not everyone in India is an Ajath Anjanappa. Not everyone has money to dine on steaks in Western-style restaurants. In India, almost 70 percent of the population lives on less than $2 a day, while a Suguna chicken breast costs about $2.50 per pound. In China, millions can't afford to dine on KFC chicken, either. But this destitute people look up to the rich and take note of their growing appetite for meat. They see the butcher shops opening, the steak houses luring; they see the non-veg Bollywood stars flexing their muscles on TV. And they want meat, too.

But our planet simply can't afford Asia's hunger for meat. It can't afford the antibiotics loaded into livestock, the water that needs to be pumped into production. It can't afford the global warming that it is causing. Likewise, it can't afford the West's meat addiction, either. It is time for nutrition transition, stage 5: behavioral change. But is it likely that we will markedly cut down our meat consumption in the near future? And how exactly *can* we change?

12

The Future of Our Meat-Based Diets

Imagine a world where everyone eats as much meat as Americans do today. Imagine a world where plates are full of burgers, overflowing with steaks. Imagine a world covered in hog farms and vast chicken houses. A world without land to grow much else, without enough water. Add to that image another Earth-like planet where we could ship some of the cows and pigs and chickens, or at least where we could harvest the grains needed to fill their stomachs. Without that extra planet, we simply won't be able to pull off a meaty future. Already, 33 percent of the world's arable land is being used to grow feed for livestock. If the 9.3 billion people that will likely be here in 2050 all want an American diet, we would need almost 4.5 times more meat than we produced in 2014 and about as much more milk for all the cheese and butter and ice cream. Although it's an oversimplification, try multiplying 33 percent by 4.5 and see how much arable land we would require to satisfy our hunger for animal protein. Yes, we are a bit of a planet short here. Of course, long before we run out of earth to grow meat, prices will have skyrocketed and we will have damaged our environment and our health.

Although in all likelihood by 2050 developing countries won't be able to afford as much meat as Westerners wolf down today, the fact is that appetites across the globe are growing. Even in many rich countries, per capita consumption of meat is still going up. In the early nineteenth century, the average person ate 22 pounds of meat per year. In 2013 it was 95 pounds. If the current growth rate continues, by 2050 it's likely going to be 115 pounds. Add population growth to that picture, and we will have to somehow double our meat production. In general, to feed the world in 2050, we need to increase our food production by 70 percent. Investing in meat is not the way to do it. Animals are not efficient converters of feed into food; they waste it on living. To grow a pound of flesh, a cow has to eat about thirteen pounds of cereal. In the US, livestock already gobbles 60 percent of all harvested grain.

Water is a problem, too. Animal agriculture is very water intensive. A pound of beef requires about 1,860 gallons of the precious liquid to produce. Meanwhile, the planet is drying up. We are overpumping to irrigate the fields and depleting aquifers. It's happening everywhere, including the US. In about a decade, over half of Earth's population will have to deal with water shortages. As taps go dry, wars may start over dams on international rivers, and governments may collapse.

To make matters worse, meat production is closely intertwined with climate change. Of all greenhouse gases released by humans, 14.5 percent comes from livestock. If this number doesn't seem that large, consider this: it's about the same as emissions from all of transportation combined—passenger cars, trucks, ships, airplanes, and so on. And yet we are so worried about fuel economy and miles per year traveled, about eating local and flying too much. We should worry about the meat on our plates at least as much.

According to a recent report by Chatham House, the British think tank, if we want to prevent catastrophic global warming, we need to curb our meat consumption. If we do nothing about global warming, mean temperatures could rise as much as 7 degrees Celsius before the end of this century (compared to preindustrial levels). That's bad for agriculture—more deserts, less arable land, less water. In general, higher temperatures around the planet would translate into less food, including meat. More people would go hungry.

Boosting efficiency would not be enough to solve this problem. To have a good chance at preventing temperatures from rising beyond 2 degrees Celsius—already a significant change—we have to cut our meat consumption. Scientists suggest we should replace at least 75 percent of calories from meat and dairy with those from cereals and pulses. We must go flexitarian—or, as some call it, "reducetarian"—and fast. But the problem is that people around the planet are not particularly willing to do so. They don't want lentils; they want steaks. To help solve the problem, the search for a perfect meat replacement has begun.

In the bluish hue of Riverside Studios in London, perched on a stool by a simple counter, Hanni Rützler is chewing a burger. She chews slowly, self-consciously. Finally, she looks up at the chef, nods in approval. "It's very close to meat," she says. "I was expecting the texture to be more soft. And it's not that juicy. But it's meat to me."

Rützler, an Austrian nutritional scientist and food trend expert, was one of three people chosen to taste the world's first lab-grown beef in the summer of 2013. As they ate, I and the other invited journalists strained our noses to catch a whiff of the in vitro meat (it smelled like any other burger). The whole event was widely publicized by the media across the globe. After all, it was not just the first burger grown in a petri dish but probably the most expensive one, too—it cost a staggering $330,000 per five ounces to produce.

The in vitro meat lab at Maastricht University, the Netherlands, is very different from what I had imagined. It was here that the London burger was invented and produced by a physiology professor, Mark Post, and his team. In my mind, the lab was grand: a large, industrial-like space, filled with test tubes, microscopes, and flasks, with plenty of scientists working in hushed concentration, shuffling around in their white coats. But on the day of my visit, I follow Anon van Essen, a jeans-clad lab technician, into a small room, maybe 110 square feet in area. There are only a few microscopes around, some empty boxes, and a few abandoned flasks. No people, and no meat in sight. "That's it?" I can't help but ask. Van Essen smiles. "We get that a lot," he says in his lispy Dutch accent. "Many TV crews actually filmed in another lab, pretending it was where we grow the meat, because this room was too tiny to get a decent shot." My next question is also pretty obvious: "Where is the meat? Can I see it?" Van Essen points at two big fridge-like devices by the back wall. Incubators, he explains. Inside, on fridge-like shelves, are petri dishes filled with reddish goo.

The goo ("growth medium," as van Essen calls it) is filled with satellite cells, a type of stem cell that is responsible for muscle regeneration after injury (for example, when you cut your finger, that's what repairs the muscle). Basically, van Essen tells me, the process of growing meat goes like this: Every few weeks a small slab of beef arrives at the Maastricht lab. The technicians fish out satellite cells from the muscle and place them in petri dishes in a mixture of nutrients that helps the cells multiply. Then off they go into the incubator where the cells grow into thin, 0.02-inch strands of muscle fiber. "You see it?" Van Essen lifts up one of the petri dishes and points to a grayish shadow floating inside, so minuscule I'm barely able to spot it. It's hard to imagine that you need about twenty thousand such fibers—thirty billion cells—to create a single burger. And yet the mastermind behind in vitro beef, Mark Post, believes that in ten to twenty years we may see lab-grown meat in the supermarkets. If the plans pan out, it may be quite a meat, too. Since in the future scientists will be able to obtain satellite cells through biopsy, without killing the animal, and because the process offers lots of flexibility, we could have burgers from almost any species imaginable, including endangered and extinct ones. The recently published *The In Vitro Meat Cookbook* suggests dodo wings, panda-flavored ice cream, and meat shaped like flowers. Meat could even come in the form of yarn, so that you could knit your own protein scarf—and eat it.

In theory, lab-grown meat could help solve several problems surrounding conventional meat. It could cut greenhouse gas emissions by 80 percent and water use by 90 percent. Produced in sterile labs instead of bloody slaughterhouses, it would be safer, bacteria-wise. What's more, it could be designed to have more unsaturated fats and reduced heme-iron content to stave off heart disease. But there are serious challenges, too. First and foremost, $330,000 per five ounces is a bit pricey to compete on the shelves of Walmart. Although scientists like van Essen are working hard to improve the process, cultured meat costs a lot because they still don't know how to make the cells grow fast enough and because the medium on which the cells feed is expensive. Second, there is the "yuck" factor. Even though cultured meat could be as delicious as any conventional beef or chicken, 80 percent of Americans claim that they could not swallow a piece of meat that was grown in a lab. "Franken-meat," they call it. But van Essen insists that such concerns are unfounded. "These cells are dead, like in any meat," he tells me, as he closes the incubators and leads me out of the lab. "Stem cells are everywhere: in your muscles, in your regular food. Nothing to fear."

Sometimes disgust is hard to overcome, though. I was certainly full of it the first time I put a dead cricket into my mouth. It was grown rather conventionally on an insect farm, not in a lab, but its looks (tiny eyes, blade-like wings) made me anxious. *What if it makes me sick?* I wondered. *What if I spit the thing out and embarrass myself?*

I am sitting in a Parisian bar tucked on the slopes of Montmartre with two young entrepreneurs who have launched a successful business selling insects as snacks. Clément Scellier and Bastien Rabastens, the founders of Jimini's, have no connections to the Montmartre bar—they just wanted to check out the competition. Their company offers products such as garlic-and-herb mealworms and tomato-and-pepper grasshoppers, which are sold through upscale French delis. For now, they tell me, they are trying to attract the Indiana Joneses of the world, adventurous trendsetters who will try eating bugs out of curiosity, learn to like the taste, and sell others on the idea. "I think people are not ready yet for a full dinner of insects," Scellier says. "But if you introduce bugs as an appetizer, encourage people to just take one or two—that's much more likely to happen."

As we talk, a plate of insects arrives from the kitchen. They are brown and shriveled, their bulging eyes staring at me emptily. I swallow hard and pierce one of the crickets with my fork and slide it into my mouth. Once on my tongue, the thing collapses into greasy ash. I chew and chew, the wings scratching the insides of my cheeks. I certainly wouldn't like to repeat the experience.

To my surprise, Scellier and Rabastens seem as disgusted as I feel. "That's just bad quality," Rabastens scoffs. "They are spoiling the market." The problem is, he explains, that there are not enough insect farms in the West to meet the demand, so most bugs are shipped from

Thailand, and for safety, they are dehydrated before export. In the process, all flavor and texture are lost. If you cook a piece of meat like that, it's going to taste bad. You may never give insects a second chance, even though that's like giving up on all beef after eating one burnt steak. Westerners don't allow insects much benefit of the doubt. We find most of them repulsive and would rather swat them with a shoe than sauté them in butter. Yet two billion other humans in 113 countries have no problem eating bugs.

Around the globe as many as two thousand different insect species are eaten, including honeybees, hornets, dung beetles, ants, flies, and silkworms. Many are considered delicacies. In Uganda, a pound of grasshoppers sells for more than a pound of beef. Insects not only can taste good, but also offer great nutrition. Plenty of species have very high iron and zinc content and are superior sources of protein even compared to pork or chicken. The planet could profit from a switch to an insect-based diet, too. For several years now, the Food and Agriculture Organization (FAO) has been pushing insects as one of the solutions for the looming food crisis. Insects are ultraefficient animal protein makers. Because they are cold blooded, they don't "waste" energy heating their own bodies—which is a major reason why, for example, to get a pound of protein from beef you need twelve times more feed than for crickets. What's more, we could grow insects in our own homes, almost like sprouting herbs on a windowsill—in a growth reactor or an insectarium.

So why aren't Americans and Europeans eating bugs? Arnold van Huis, professor of entomology at Wageningen University, the Netherlands, and a world-class expert on edible insects, believes that the West's dislike of edible insects is rooted in economics. In temperate climates, he told me, it is more difficult to collect enough bugs to eat than it is in other parts of the planet. Because of their cold bloodedness, in the tropics insects grow bigger (more food per creepy-crawly), and they tend to clump together, such as during locust plagues, making harvests easier. It simply didn't make much sense for our European ancestors to collect bugs. To boot, industrialization cut the West off from nature—and so we started to demonize insects.

But recently bugs are beginning to crawl into Western cuisines. Cricket energy bars are selling in the US, a British company is developing sushi-like "ento" boxes, and in the Netherlands top chefs create recipes for such dishes as mealworm quiche. Already, one out of five European meat eaters claims to be ready to eat insects. How to convince the rest?

One way would be to hide the bugs—and from a certain perspective, we've been eating them this way for years. By law, particles of insects are allowed in other products. In the States, 250 ml of canned citrus juices can contain five or more fly eggs, and peanut butter up to thirty insect fragments. Also, a red dye made from cochineal beetles is used in many foods. To start Westerners on the new diet, we could add insects to regular meat products—for instance, make meatballs that are 30 percent mealworms. We could also use insects as feed for livestock or make insect flour and 3-D print it into visually appealing products—some British scientists are doing this already. But just as with in vitro meat, there are many challenges that face the brave new world of insect eating. The costs are still too high, the laws inadequate, and the research lacking. And as animal rights advocates point out, insects are animals, too, and may perceive pain. However, in the West, insects may soon follow the path of sushi and become a fashion. Will they feed the world? Probably not—at least not by themselves. But they could certainly help wean us off vertebrate meat.

Even though a hornet burger still hails from a rather distant future, there are plenty of fake meats available in the here and now. As opposed to lab-grown meat or edible bugs, these contain no animal protein whatsoever. I'm talking about all the tofurkies, meatless meatballs, and other mock meats of the world. Some of them are horrible—chewy, tasteless, and definitely not "meaty." Others are so good that even the best of chefs have trouble believing they are not the real thing. To try one of those, I enter a small butcher store in downtown Hague (yes, the Netherlands again—they really are the leaders in the search for meat alternatives). In many ways, that Hague butcher store is quite old school. There is a vintage scale and a manual meat grinder standing on the counter, with a few meat choppers scattered around. The only thing that's missing is the meat itself. Welcome to the Vegetarian Butcher.

Jaap Korteweg, big, bald, and smiley, greets me at the door. Korteweg, the "butcher" himself, used to be a meat-loving farmer but became disillusioned by modern livestock production and turned vegetarian. He had a problem, though. He still loved meat. He started dreaming of building a stainless-steel cow into which one could pour grain and get meat in return, and so he founded the Vegetarian Butcher.

Of course, the process in which Korteweg's fake meat is produced is far more complicated and less romantic than a stainless-steel cow churning out steaks. Like other fake meats of the so-called third generation (new, better), Korteweg's sausages and burgers are made by breaking and reassembling protein molecules taken from plants such as soybeans, peas, or lupin (a type of legume). The actual forming of fake meat happens in machines similar to those used to make spaghetti and breakfast cereal. Basically, you apply heat and moisture to a protein mix, knead it into a dough, then push it through a special die—and out come chunks of "chicken" or "tuna." What's tricky is getting the flavor right. How do you imitate, for instance, the taste of canned tuna? Korteweg tells me that you need fermented yeast, some seaweed, and three different plants (one of them is wheat, but he doesn't want to give up the others—it's a trade secret). For beef, on the other hand, you need onions, carrots, and yellow peas. "You make beef out of carrots?" I ask, doubt resonating in my voice. "Come and taste it for yourself," Korteweg replies and calls me to the back of his store to the kitchen: a place of sizzling pans and clattering pots, with charred, fatty smells hanging in the air.

To tell you the truth, I still can't quite believe that what I ate in Korteweg's kitchen was not meat. I tried his chicken, his beef, his canned tuna, and the illusion was perfect. I'm sure in the past I dined on real chicken that was less chickeny than that mock-vegetarian one. It was succulent, resistant to the teeth just enough, and rich in flavors. And I wasn't the only one who had trouble telling the real from the fake. When Ferran Adrià, one of the best chefs in the world, took a bite of the Vegetarian Butcher's chicken, he couldn't believe it was made of plants. He guessed it to be a chicken thigh from the south of France.

It was only very recently that the improved availability of vegetable proteins, as well as other technological advancements, has made it possible to truly imitate meat. It's not just the Vegetarian Butcher that excels at this. Other companies, like Beyond Meat in the US, are doing a great job, too. When in 2013 one Whole Foods store mistakenly switched labels on salads containing real chicken and Beyond Meat's fake one, no customers noticed the difference and complained. What's more, nutritionally veggie meats can be as complete in protein as meat, yet much leaner. So why don't we all just dig in?

The problem is fake meats still have a reputation for tasting horrible. In one episode of the popular TV show *Breaking Bad*, one of the characters, Walter Junior, doesn't want to eat the veggie bacon his mother has cooked. "This smells like Band-Aids," he says. "I want real bacon. Not this fake crap." Just as with eating insects, one bad experience with no-chicken wings can scar one for life. But the good news for the planet and for the future of our food supply is that fake meat consumption is going up. The global meat substitutes market, which was worth over $3 billion in 2013, is expected to grow 45 percent by 2019. Compared to real meat, though, these numbers are still very small. In 2011, sales of meat substitutes in the US totaled just 0.2 percent of what Americans paid to get their meat. Meanwhile, scientists and marketers alike are coming up with ways to convince Westerners to swap hot dogs for "not dogs." Saying it tastes and cooks "just like meat" works pretty well. So does suggesting easy recipes. According to Ingrid Newkirk, the president of PETA, the single best way to convince people to eat fake meat is to have them try it. "We make vegan ham sandwiches and give them away," she told me. "People are amazed. They say: 'This tastes really good. Are you sure this is made from plants?'"

If the idea of dining on fake chicken or beef repels you, think about it: most of us eat meat substitutes already. If you consume sausages and other processed meats, meat-lover pizzas and ready-to-eat meals, you consume quite a lot of soy protein, which is often used to extend the meat in such products. In the States, up to 30 percent of meat served in the National School Lunch Program is actually soy protein. Some scientists suggest that simply extending most of our meat could help reach climate change goals. Just mix more soy and lentils into burgers, and we will all be better off. Yet the conquest of supermarket shelves by meat replacements likely won't turn the whole planet vegetarian all by itself. It could help tremendously, of course, but if humanity is to significantly reduce meat consumption—and it should—more incentives and solutions are needed.

The first would be to simply waste less. In North America and Europe, over 20 percent of meat is thrown out—either because it wasn't up to the meat producer's standards, it didn't sell at the market, or it didn't get eaten at home. Half of unused meat is tossed into the garbage by consumers themselves. This means many people can cut down their meat consumption without even changing their diet simply by improving shopping habits, planning meals better, and learning how to use leftovers or freezing them for future use. This would also save money—money that may be needed for a growing butcher's bill, if a meat tax is introduced.

The meat tax is a second solution that may push the Western world toward the fifth stage of nutrition transition (behavioral change) and a more vegetarian future. Scientists, journalists, and politicians alike are already calling both for a halt on meat industry subsidies and for a tax (similar to the cigarette tax), which would increase the price of beef, pork, and chicken. Studies confirm that this could be an effective way of discouraging the consumption of animal protein. In Europe, it has been calculated that turning the tap off on agriculture subsidies for dairy and meat would save at least thirteen thousand people a year from dying of stroke and heart disease. And that's a conservative estimate. It may be a challenge, though, to introduce meat taxation. Just look at what happened recently in Denmark. A "fat tax" was proposed there, amounting to 0.77 euro per pound of saturated fat found in food products, including meat. The livestock industry was not happy. They wrote letters to the government, threatened lawsuits, and basically did everything they could to stop the tax. In the end, they succeeded—the fat tax idea was abandoned.

The third important solution, a cultural shift that holds the potential to significantly curb our meat consumption, is promoting and rewarding flexitarianism or, perhaps more accurately, "reducetarianism." Many devout vegetarians probably won't like what I'm going to write next, but I'm not alone in thinking that giving more kudos to people for even slightly changing their eating habits is important. One who concurs is Peter Singer, considered by many "the most influential living philosopher"—famous for challenging traditional notions of applied ethics and for his best-selling book *Animal Liberation*.

I called Singer to ask him about his views on the likelihood that humanity will ever go vegetarian and what would need to change for that to happen. In a soft voice that made him sound much younger than he really is, Singer told me that besides a meat tax, of which he is a big proponent, deradicalizing vegetarianism would encourage more people to try it, even just part-time. "We should stop abusing people if they are not completely vegan or vegetarian," he explains. "If you announce that once you become vegetarian you must starve to death rather than have any meat pass your lips, people are going to say: 'That's crazy, I'm not going to do that.' If we want the majority to reduce meat consumption, I don't think insisting on absolute dietary purity is the way to get there." This approach allows people to try vegetarianism or veganism without needing to jump headfirst into a huge well of ideology and restrictions. Singer himself says he is a "flexible vegan": he tries to avoid animal products, but when a situation makes it too difficult (visiting friends, travel), he won't always turn down a dish just because it contains cheese or eggs.

The risk, of course, is that encouraging reducetarianism instead of outright vegetarianism or veganism will mean effectively letting people continue business as usual. Maybe even some of the current vegetarians would go back to eating meat. Maybe. Most likely, though, making a plant-based diet less of an all-or-nothing proposition and less attached to a whole set of lifestyle choices will encourage people to try it. As Paul Rozin once told me: "People get credit for sorting their recycling even if they don't always do it 100 percent. But we don't get any credit from either vegetarians or nonvegetarians for cutting down on meat." Maybe it's time we should. You could try Meatless Mondays, VB6 (vegan before 6:00 p.m.), or Veganuary (going vegan in January), or you could basically make up your own scheme—even "vegetarian on every second

Wednesday of the month, as long as it's not raining." At the time of writing, Meatless Mondays were happening in twenty-nine countries across the globe. In the States, 39 percent of people were cutting down on meat (mostly for health reasons). In Germany that number was 41 percent. It would be good if reducetarians got more recognition for their efforts, a badge to wear, either literal ("I reduced 10 percent!") or simply figurative. Something to be proud of.

Moreover, polls show that once we start banishing animal products from our plates, we keep adding to the list of things we don't eat: first goes red meat, then chicken, then fish, then milk and eggs. Does this mean that by following the flexitarian/reducetarian path one day the world will actually end up vegetarian? The answer depends on whom you ask. When I posed this question to Newkirk, she said yes. So did Singer and Korteweg. Lord Stern, former vice president of the World Bank and adviser to the UK government, believes that in the future eating meat will become as socially unacceptable as drinking and driving. But Morgaine Gaye, a British food futurologist, thinks that humanity won't give up meat completely in the foreseeable time.

I'm meeting Gaye in a rather nonfuturistic venue: Charlotte Street Hotel in London, a boutique place with a lot of old-school charm. Gaye herself, though, looks quite the part. She has a short, pointy hairstyle that suggests her "futuristic" outlook (although to be fair, based on looks alone, a quarter of London's population could be considered futurologists). Gaye's job involves consulting food companies on the future of our eating habits. When it comes to food, she knows what's "in," what's "out," and where we are heading. And she doesn't think meat is destined to be "out" quite yet, at least not completely. "I don't think we will necessarily go vegetarian, but I think people will eat a *lot* less meat," she says. "As meat becomes less affordable, we will go back to considering it a treat. I think people will value meat even more than they do now. What makes things fashionable is unavailability, and meat will be hard to get. It will be expensive. We will see high-end butcher shops appearing—it's already happening." Gaye tells me about Victor Churchill, a luxury meat "boutique" that opened recently in Sydney, where cuts of beef and pork are displayed as if they were Louis Vuitton handbags. She also tells me that in the future, there will likely be more meat snacks. Things to grab and go. Basically, it's bye-bye twelve-ounce steaks, hello tiny morsels of very expensive meats.

What's more, such meat snacks could be lab grown, too. Or "cultured," as Andras Forgacs prefers to call them. Forgacs's company, Modern Meadow, which is backed by PayPal's billionaire Peter Thiel, is working on creating "cultured" concept meats rather than growing burgers in petri dishes. As Forgacs once told me: "Why focus so much on the 'I can't believe it's not slaughtered meat,' if we can have something delicious, healthier, safer, more nutritious, more convenient?" That's why Modern Meadow is now developing cultured "steak chips"—something between a potato chip and beef jerky that's "super-healthy and has lots of protein" (in Forgacs's words). Modern Meadow's "steak chips" could be on the supermarket shelves in a few years. And if the trend for meat snacks sets in, they may be quite a hit.

Now let's imagine that the world will, at some point, actually go completely vegetarian. What would happen then? According to some, that future would be quite bleak. Think unemployment, economy in ruins. Cows extinct. Pigs extinct. Bland cuisine. Is that a likely scenario? Thankfully not. How would, then, a meatless Earth differ from the one we have now?

To begin with, we could well spend less time on the toilet—struggling with food poisoning—and live longer. A recent USDA study showed, for example, that a quarter of chicken breasts have *Salmonella* bacteria on them, and 21 percent are positive for *Campylobacter*. Yes, proper handling and cooking help prevent food poisoning, but "proper" is often not what happens in the reality of our kitchens. According to the CDC, meat and poultry are the most common food sources of fatal infections—a lot of them caused by *Salmonella* and *Listeria* bacteria. Of course, in a vegetarian scenario, we would also likely live longer because, as studies show, eating meat may increase the risk of cancer, cardiovascular disease (CVD), diabetes, and so on. As authors of one study concluded, "red meat consumption is associated with an increased risk of total, CVD, and cancer mortality."

In a meatless world, we would also have fewer reasons to worry about antibiotic resistance. In the US, 80 percent of the total volume of antibiotics are used in livestock production. At the same time, twenty-three thousand Americans die every year from antibiotic-resistant bacteria. There may be a connection between the overuse of antibiotics in animal production and bacterial resistance to antibiotics. Studies have found, for example, that in experimental swine herds regularly exposed to antibiotics, *E. coli* bacteria were much more resistant than in herds that had not been exposed to antibiotics for a long time.

It could be easier to breathe, too. In Europe scientists calculated that cutting meat consumption by half would lower total nitrogen emissions by 40 percent, which, as a 2014 study predicts, would "result in a significant improvement in both air and water quality in the EU." Though pastures would disappear, wildlife would return. In Europe, 23 percent less cropland per person would be needed to produce food. A similar thing would happen in the States. Yes, we would have far fewer cows, pigs, and chickens, but instead we would have space for more wild animals. As Jeremy Rifkin wrote in *Beyond Beef*: "Millions of creatures, many of whom have inhabited this earth for millennia, will regroup, reproduce, and repopulate the forests." Sure, not all of the new acres would be returned to a natural state. Some of the postlivestock land would be likely used to grow bioenergy crops, such as switchgrass and willow. Again, the skies would be cleaner for that.

What about unemployment, though? Worldwide, the livestock industry amounts to about 1.3 billion jobs. If the whole planet went vegan, all those jobs would be gone. If we went vegetarian, quite a few would be left, of course, but the impact would still be large. The majority of Earth's farmers live in poor countries, grazing animals near their homes. Yet the livelihood of these people is also at risk if the rich of the world overindulge in meat—because of climate change and because of the multinational meat corporations pushing out small producers. To take just one example: the European broiler industry has just about destroyed local chicken production in parts of West Africa. At least in the vegetarian scenario, we have a much higher chance of winning the climate-change battle, saving land in developing countries from flooding and desertification, and reducing heat, which destroys crops. Vegetarian humanity would need less livestock, yes, but it would need more veggie burgers, tofu, lentils, and greens. After all, Jaap Korteweg used to be a livestock farmer and now successfully makes his living producing mock meat.

It is true, though, that for some people going vegetarian, or even cutting their meat consumption, may be against their best interests. I'm talking about those inhabitants of Earth who live in poverty, grazing their animals on roadsides and on city garbage (in Havana, sixty-three thousand pigs live on the streets), and those who live on land that is too dry, steep, or hot for growing crops. These people can't afford a plant-based diet. If you live in Uganda and all you have is plantain, exchanging it for a scrawny chicken to satisfy your protein hunger may be, after all, a very good idea.

But even if there were no lack of meat alternatives and no worries about unemployment in a postmeat world, it would still be hard for humanity to completely let go of meat. The challenge facing makers and marketers of edible insects, lab-grown meat, and veggie burgers alike is that they have to replace not only the taste of animal flesh and its nutrition but also all the symbolism that it carries. We crave meat because it stands for wealth and for power over other humans and nature. We relish meat because history has taught us to think of vegetarians as weaklings, weirdos, and prudes and because the meat industry knows how to sell its products. We die for meat (sometimes literally) because the mistakes of nineteenth- and early twentieth-century science led us to believe in the protein myth. That's a lot of hooks to let go of if humanity is ever to go vegetarian or even just significantly reduce its meat consumption.

This isn't to say that giving up meat is impossible for humanity, but the road will be long and bumpy. Meat substitutes have to become prestigious (celebrity endorsement helps); they have to be cheap but not too cheap, so they can still stand for "I've made it in life"; they have to become equated with masculinity (ads featuring bodybuilders, perhaps?); and they have to become visible and omnipresent, so that we get used to them and develop new habits. History has shown that people do acquire appetites for foods that were long considered disgusting or inferior. Potatoes used to be thought of as pig food in Europe, dangerous to human health. It took the French court a lot of effort to convince people to eat them, in order to improve diets and fight hunger. Marie Antoinette wore potato flowers in her corsage, and a potato field by the palace was ostensibly guarded to spur curiosity (it worked—the moment the guards were called off, peasants stole the whole crop). Tomatoes were looked upon with suspicion well into the eighteenth century, lobster used to be poor people's grub, and even pizza didn't catch on right away in the States. Communist fare, people called it. Meanwhile, foods can lose their prestige, too. Take what happened with white bread. For centuries it used to mean status and wealth; nowadays it's considered inferior to whole grain.

It's hard to imagine humanity going full veg in the very near future. Thousands of years of our long love affair with meat must be undone. If meat becomes unavailable again, people will crave it. That's simply the way we operate. In Margaret Atwood's dystopia *Oryx and Crake*, the world survives on lab-grown meat, while the real stuff is as rare as diamonds and so ferociously desired. That is one possible future.

What is likely to happen, though, is that meat substitutes will sneak up on us, the way potatoes and pizzas did. As long as they taste good (and some of them truly do), and as long as they gain popularity and status, mock meats may eventually replace quite a lot of "real" meat. So far, plant-based "veggie" meats are the cheapest and the least disgust inducing of meat substitutes, and the process of their production is the most advanced. Still, edible insects and cultured meat may also play a role in reducing our meat consumption in the future, and once the idea catches on, it may quickly gain traction.

Epilogue

The Nutrition Transition, Stage 5

Why do we eat meat? At the end of the day, despite the many complex reasons rooted in evolution, history, and culture, the most basic answer would be: because we can. We are omnivores, and meat is food—a food loaded with amino acids, which satisfy our protein hunger, and fat, which provides us with energy. For millennia, meat kept us fed. It helped us grow our big brains and move out of Africa. In a way, it made us human.

Yet meat is no longer as good for us as it used to be. With abundant plant-based foods, we, in the developed world, don't actually need animal flesh for its nutrients. What's more, numerous studies show that meat consumption may lead to cancer, diabetes, and heart disease. We also don't have enough planet to feed all humans the Western, carnivorous diets they'd prefer, given the chance. If we don't reduce our meat consumption significantly, we are far more likely to face global warming, water shortages, and pollution.

It's time for the next and final stage of nutrition transition—behavioral change. It's time to switch from meat toward a diet based on vegetables, grains, fruits, and legumes, such as beans and lentils. It may seem daunting, but we've done things like this in the past. Several times before, when Earth's climate changed, our ancestors adjusted their eating habits. A few million years ago *Purgatorius* took advantage of the new wealth of fruits. Later, when the climate cooled and plant foods became harder to find, early hominins started to eat meat. When in India consuming cows became uneconomical, they made it taboo. When there was not enough land to grow animal protein in medieval Japan, their rulers banned many meats.

We may not need to prohibit carnivorous diets, but as the climate and economy change once again, so should our eating habits. It won't be easy, of course. Meat is not just a pleasure to the taste buds but also a symbol-laden element of our culture. It stands for wealth, for masculinity, and for power over the poor and over nature. For many in the developing world, it symbolizes modernity, progress, and a decisive break with traditional, hierarchical society.

To enter the final, fifth stage of the nutrition transition, we should first become aware of meat's many meanings—only then can the hooks be released one by one. The taste of meat can be replaced by products containing meat's potent mixture of umami, fat, and the aromas created by the Maillard reaction. Protein hunger can be satisfied with lentils, beans, and even peanut butter sandwiches. Government policies can be reversed, with subsidies diverted and a meat tax introduced. Ag-gag laws and food disparagement laws can be changed. We can stop propagating protein myths and make meat replacements widely available, so they can turn into a habit themselves. By becoming aware of our eating scripts (summer evening equals grilling equals beef burgers), we can choose to change them. We can take advantage of our psychological wiring and create positive associations of vegetarian meals by pairing them with foods we already love (a veggie dinner followed by ice cream) and by eating them at fun, social occasions. We should emphasize that plant-based diets are convenient and economical and not just healthy, because that's what drives people's buying choices. We should try to change the image of a vegetarian diet by showing athletic, masculine men eating a plant-based diet, emphasizing that "veg" can make you strong and beautiful. That is, after all, how meat has been sold to us for years.

Don't get me wrong: I'm not saying we should all turn vegetarian tomorrow. Even though I do believe that in the future humanity will eat mostly plant-based foods, I also believe that pushing for dietary purity is not the way to go, and it may actually backfire—as it has a few times in the past. Instead, we should reward cutting down meat consumption—whether we call it reducetarianism, flexitarianism, or stage five-ism and whether we cut down by 5 percent or 99 percent. Strict vegetarians and vegans should stop criticizing vegetarians who sometimes secretly eat meat. After all, compared to the Western average, they likely did manage to change their diets substantially. Building barricades between vegetarians and meat eaters didn't work in the past, and there is no reason why it should now. In a similar vein, instead of always fighting the meat industry as evil incarnate, it may sometimes make sense to work *with* it—the way Temple Grandin does or the way the Vegetarian Butcher does, who offers his products in conventional meat shops. If your goal is to improve your health, limit animal suffering, and increase our chances to combat climate change, it may actually be better to eat a bit of meat now and then and a lot of plant foods, rather than be a strict lacto-ovo vegetarian on a diet loaded with cheese, milk, and eggs. It's just a matter of adding the numbers. If you are an ethical vegetarian, think about it: Which saves more lives—one person stopping eating meat altogether or millions cutting out just one meat-based meal a month? The same goes for meeting the climate-change goals, of course. Yes, it would be better if millions went vegetarian, but this won't happen overnight. The meat hooks are far too strong for that.

What's important is to be aware of factors that drive our food choices, instead of blindly following our routines, our culture, and advertising. If we are to progress to the fifth stage of the nutrition transition, that, I believe, is the first step.

Acknowledgments

It takes a village to write a book, and I wouldn't have been able to write *Meathooked* if it wasn't for the help of my "village" (which happens to stretch across the globe). I'm particularly grateful to all the scientists who helped me understand the nuances of their academic fields and who commented on my work: Carol J. Adams, Linda Bartoshuk, Brock Bastian, Grégory Bayle, Gary Beauchamp, Paul Breslin, Henry Bunn, T. Colin Campbell, Temple Grandin, Arnold van Huis, Gáspár Jékely, Edward Mills, Stéphane Péan, Stephen Simpson, and Richard Wrangham. Thank you Mark Post and Anon van Essen: seeing meat grow in your Maastricht lab was an unforgettable experience. Briana Pobiner: I never would have thought I'd hold a million-year-old elephant bone in my hand—thank you. For their scholarly advice, I'm also indebted to Leslie C. Aiello, Nikos Alexandratos, Neal Barnard, Adam Drewnowski, Robert Eisenman, Daniel Fessler, Hal Herzog, Urban Jonsson, R. S. Khare, Joe Millward, Marion Nestle, Chris Otter, David Penny, Anthony Podberscek, Rashmi Sinha, Martin Smith, Erik Sperling, Brian Wansink, Yunxiang Yan, and Ryan Zarychanski. Over the years that it took to write this book, I've met many fascinating people, whom I wanted to thank for letting me catch a glimpse of their meat-related worlds and for the opportunity to discuss the past and the future of humanity's meat addiction: Andras Forgacs, Morgaine Gaye, Ifor Humphreys, Kate Jacoby, and Richard Landau (and big thanks for the delicious food), Scott Jurek, Evelyn Kimber, Niko Koffeman, Jaap Korteweg, and Paul Bom (please bring Vegetarian Butcher to France),

Kristin Lajeunesse, Howard Lyman, Ingrid Newkirk, Bastien Rabastens, Bill Roenigk, Hanni Rützler, Clément Scellier, Peter Singer, and Ajath Anjanappa—thank you for being my guide to the food scene of Bengaluru.

For relentlessly honing my science-writing skills (by making me cut and edit and cut some more), I'd like to thank Pooh Shapiro of the *Washington Post*. Also a big "thank-you" to all the editors at *Polityka*, *Gazeta Wyborcza*, the *Boston Globe*, the *Los Angeles Times*, *Scientific American*, *New Scientist*, and *The Atlantic*: over the years, you have helped me become a better writer. And in particular I wanted to thank Stephen Northfield, the former foreign editor of *The Globe and Mail*, who back in 2009 took a chance on me and assigned me my first-ever feature article in English.

This book wouldn't have happened if it wasn't for my amazing agent Martha Magor Webb: thank you for all your hard work, for all the advice and encouragement. Alex Littlefield: your enthusiasm made it so much easier for me to research and write *Meathooked*—thank you for being such a great editor. For all their hard work on the manuscript and for their helpful comments, I want to thank my editors Dan Gerstle and Brandon Proia: you've made this book so much better. Katie Haigler, Kate Mueller, Melissa Raymond, Melissa Veronesi, and the whole team at Basic Books—without you *Meathooked* wouldn't exist.

A big thank-you goes also to my family, and to my mom in particular, for helping take care of my baby daughter so that I could write (and so that I could go on holidays and rest after all the writing). To my dad, for insisting that I learn English. To my mother-in-law, for coming over to help in times of need. To my friends, who over board games and in sandboxes helped me take a break from the fascinating but challenging world of meat eating. Last but definitely not least, to my husband, Maciej, who has helped me believe that, yes, writing is a proper career, and who has relentlessly supported me over the years (and who has patiently dealt with all the writer drama). And to my daughter, Ellen—because she is a ray of sunshine that can brighten the cloudiest of days.

Notes

Introduction

1 **The data it quoted:** Rashmi Sinha et al., "Meat Intake and Mortality: A Prospective Study of Over Half a Million People," *Archives of Internal Medicine* 169 (2009): 562–571.

1 **She didn't want to clog:** Ibid.

2 **I don't hover:** Jaroslaw Adamowski, "Polish Meat Consumption to Expand in 2015," accessed December 2, 2014, www.globalmeatnews.com/Industry-Markets/Polish-meat-consumption-to-expand-in-2015.

3 **Americans devour:** Mark C. Eisler et al., "Agriculture: Steps to Sustainable Livestock," *Nature* 507 (2014): 33.

3 **In the meantime:** Hope R. Ferdowsian and Neal D. Barnard, "Effects of Plant-Based Diets on Plasma Lipids," *American Journal of Cardiology* 104 (2009): 947–956.

3 **According to studies:** Denis E. Corpet, "Red Meat and Colon Cancer: Should We Become Vegetarians, or Can We Make Meat Safer?," *Meat Science* 89 (2011): 310–316.

3 **High intake of red meat:** Astrid Steinbrecher et al., "Meat Consumption and Risk of Type 2 Diabetes: The Multiethnic Cohort," *Public Health Nutrition* 14 (2011): 568–574.

3 **In one widely cited study:** An Pan et al., "Red Meat Consumption and Mortality," *Archives of Internal Medicine* 172 (2012).

3 **Meanwhile, studies show:** Gary E. Fraser and David J. Shavlik, "Ten Years of Life," *Archives of Internal Medicine* 161 (2001): 1645–1652.

3 **American meat consumption:** "Meat Consumption in the United States, 1909–2012," Earth Policy Institute, accessed December 2, 2014, www.earth-policy.org/datacenter/xls/book_fpep_ch3_14.xlsx.

3 **The OECD estimates:** "OECD-FAO Agricultural Outlook 2011–2020," OECD Publishing and FAO, accessed December 2, 2014, http://dx.doi.org/10.1787/agr_outlook-2011-en.

3 **In China, meat consumption:** Mindi Schneider and Shefali Sharma, "China's Pork Miracle?," Institute for Agriculture and Trade Policy, February 17, 2014, accessed December 2, 2014, www.iatp.org/files/2014_03_26_PorkReport_f_web.pdf.

3 **The media have reported:** Timi Gustafson, "Fight Climate Change with a New Diet," HuffPost Living, July 2, 2014, accessed December 2, 2014, www.huffingtonpost.ca/timi-gustafson/vegetarian-diet_b_5552288.html.

3 **Producing one calorie:** "Fight Global Warming by Going Vegetarian," PETA, accessed December 2, 2014, www.peta.org/issues/animals-used-for-food/global-warming/.

4 **Meat eating is responsible:** Nathan Fiala, "How Meat Contributes to Global Warming," *Scientific American*, February 2009, accessed December 2, 2014, www.scientificamerican.com/article/the-greenhouse-hamburger/; IATA Technology Roadmap 2013, accessed December 2, 2014, www.iata.org/whatwedo/environment/Documents/technology-roadmap-2013.pdf.

4 **According to some:** James Hansen et al., "Ice Melt, Sea Level Rise and Superstorms: Evidence from Paleoclimate Data, Climate Modeling, and Modern Observations That 2C Global Warming Is Highly Dangerous," *Atmospheric Chemistry and Physics* 15 (2015): 20059–20179.

4 **According to a 2003 Gallup poll:** David W. Moore, "Public Lukewarm on Animal Rights," Gallup News Service, May 21, 2003, accessed December 2, 2014, www.gallup.com/poll/8461/public-lukewarm-animal-rights.aspx.

4 **In one study, 81 percent:** Andrew Rauch and Jeff S. Sharp, "Ohioans' Attitudes About Animal Welfare," January 2005, accessed December 2, 2014, http://ohiosurvey.osu.edu/pdf/2004_Animal_report.pdf.

4 **We crowd our egg-laying:** "The Egg Industry," PETA, accessed December 2, 2014, www.peta.org/issues/animals-used-for-food/factory-farming/chickens/egg-industry/.

5 **According to Gallup:** George Gallup, "The Gallup Poll," *Washington Post*, October 2, 1943, 9.

5 **By 2012 the number:** Frank Newport, "In U.S., 5% Consider Themselves Vegetarians," July 26, 2012, accessed December 2, 2014, www.gallup.com/poll/156215/consider-themselves-vegetarians.aspx.

5 **But another survey:** Hal Herzog, "Why Are There So Few Vegetarians?," *Psychology Today*, September 6, 2011.

6 **Is it the skillful marketing:** "U.S. Meat and Poultry Production & Consumption: An Overview," American Meat Institute, April 2009, accessed December 2, 2014, www.meatami.com/ht/a/GetDocumentAction/i/93335.

7 **I examine these hooks:** "Corn Subsidies," EWG Farm Subsidies, accessed December 2, 2014, http://farm.ewg.org/progdetail.php?fips=00000&progcode=corn.

7 **If you are one of the:** Eliza Barclay, "Why There's Less Red Meat on Many American Plates," NPR, June 27, 2012, accessed December 2, 2014, www.npr.org/blogs/thesalt/2012/06/27/155837575/why-theres-less-red-meat-served-on-many-american-plates.

CHAPTER 1: ENTER MEAT EATERS

10 **And yet, at some point:** Gáspár Jékely, "Origin of Phagotrophic Eukaryotes as Social Cheaters in Microbial Biofilms," *Biology Direct* 2 (2007).

10 **"It's like a community:** Gáspár Jékely, phone interview by author, December 17, 2013.

10 **These, though, were:** Silvester de Nooijer, Barbara R. Holland, and David Penny, "The Emergence of Predators in Early Life: There Was No Garden of Eden," *PLoS ONE* 4 (2009).

10 **They were essential:** Stefan Bengtson, "Origins and Early Evolution of Predation," *Paleontological Society Papers* 8 (2002): 289–317.

10 **With time and generations:** Jékely, "Origin of Phagotrophic Eukaryotes."

11 **Once the ancient bacteria:** Bengston, "Origins and Early Evolution."

11 **In the warm oceans:** Roy E. Plotnick, Stephen Q. Dornbos, and Junyuan Chen, "Information Landscapes and Sensory Ecology of the Cambrian Radiation," *Paleobiology* 36 (2010): 303–317.

11 ***Cloudina* was an anemone:** Ibid.

12 **Jellyfish-like animals:** Katja Seipel and Volker Schmid, "Evolution of Striated Muscle: Jellyfish and the Origin of Triploblasty," *Developmental Biology* 282 (2005): 14–26.

12 **Of course, how that:** "Off the Beaten Palate: Sea Anemone," June 7, 2012, accessed November 28, 2014, http://shanghaiist.com/2012/06/07/off_the_beaten_palate_sea_anemone.php.

12 **One of those earliest meat eaters:** Jean Vannier, "Gut Contents as Direct Indicators for Trophic Relationships in the Cambrian Marine Ecosystem," *PLoS ONE* 7 (2012).

12 **They resembled tubes:** Ibid.

12 ***Nectocaris* squirted itself:** Martin R. Smith and Jean-Bernard Caron, "Primitive Soft-Bodied Cephalopods from the Cambrian," *Nature* 465 (2010): 469–472; Martin R. Smith, phone interview by author, December 9, 2013.

13 **It was the largest meat eater:** "*Anomalocaris canadensis* (proto-arthropod)," accessed November 28, 2014, http://paleobiology.si.edu/burgess/anomalocaris.html.

13 **The whole animal kingdom:** Bengston, "Origins and Early Evolution."

14 **If Earth hadn't become:** Erik Sperling et al., "Oxygen, Ecology, and the Cambrian Radiation of Animals," *Proceedings of the National Academy of Sciences* 110 (2013): 13446–13451.

14 ***Purgatorius* was an accomplished:** Matt Kaplan, "Primates Were Always Tree-Dwellers," *Nature News* (2012).

15 **Their guts were:** Margaret J. Schoeninger et al., "Meat-Eating by the Fourth African Ape," in *Meat-Eating and Human Evolution*, eds. Craig Stanford and Henry T. Bunn (Oxford, UK: Oxford University Press, 2001), 186.

15 **Some researchers suggest:** Ibid., 191.

16 **Much of the rain forest:** Karen Lupo, "On Early Hominin Meat Eating and Carcass Acquisition Strategies: Still Relevant After All These Years?," in *Stone Tools and Fossil Bones: Debates in the Archaeology of Human Origins*, ed. Manuel Domínguez-Rodrigo (New York: Cambridge University Press, 2012), 121.

16 **Moving on two legs:** Craig B. Stanford, "A Comparison of Social Meat-Foraging by Chimpanzees and Human Foragers," in *Meat-Eating and Human Evolution*, eds. Stanford and Bunn, 137.

17 **Cut marks are V shaped:** Briana Pobiner, interview by author, Washington, DC, October 29, 2013.

18 **The oldest undisputed:** Lupo, "On Early Hominin Meat Eating," 128.

18 **Some were much, much:** Robert Blumenschine and Briana Pobiner, "Zooarchaeology and the Ecology of Oldowan Hominin Carnivory," in *Evolution of the Human Diet: The Known, the Unknown, and the Unknowable*, ed. Peter S. Ungar (Oxford, UK: Oxford University Press, 2007), 175.

18 **In the minds of our ancestors:** Ibid.

19 **Marrow from a tiny:** Ibid., 178.

20 **Without stone tools:** Nerissa Russell, *Social Zooarchaeology: Humans and Animals in Prehistory* (Cambridge, UK: Cambridge University Press, 2012), 149.

20 **Most mammals have them:** Horst Erich König and Hans-Georg Liebich, *Veterinary Anatomy of Domestic Mammals: Textbook and Colour Atlas* (Stuttgart, Germany: Schattauer, 2007), 314.

20 **Besides, human canines:** Matt Cartmill and Fred H. Smith, *The Human Lineage* (Hoboken, NJ: Wiley-Blackwell, 2009).

20 **Even though canine teeth:** Jeremiah E. Scott, "Nonsocial Influences on Canine Size in Anthropoid Primates," December 2010, accessed November 28, 2014, http://repository.asu.edu/attachments/56070/content/Scott_asu_0010E_10059.pdf.

20 **Most likely the canine teeth:** Ibid.

21 **The true meat-eating:** Patricia Smith and Eitan Tchernov, eds., *Structure, Function, and Evolution of Teeth* (London: Freund, 1992), 216.

22 **The famed British-Kenyan:** Travis R. Pickering and Manuel Domínguez-Rodrigo, "Chimpanzee Referents and the Emergence of Human Hunting," *The Open Anthropology Journal* 3 (2010): 111.

22 **Once early *Homo*:** R. Dale Guthrie, "Haak en Steek: The Tool That Allowed Hominins to Colonize the African Savanna and to Flourish There," in *Guts and Brains: An Integrative Approach to the Hominin Record*, ed. Wil Roebroeks (Leiden, Netherlands: Leiden University Press, 2007), 155–161.

22 **In the jungles of Senegal:** Travis Rayne Pickering and Henry T. Bunn, "Meat Foraging by Pleistocene African Hominins," in *Stone Tools and Fossil Bones*, ed. Domínguez-Rodrigo, 157.

24 **Lions scavenge frequently:** Craig Packer, David Scheel, and Anne E. Puset, "Why Lions Form Groups: Food Is Not Enough," *The American Naturalist* 136 (1990): 1–19.

CHAPTER 2: BIG BRAINS, SMALL GUTS, AND THE POLITICS OF MEAT

26 **There is quite a lot:** Nerissa Russell, *Social Zooarchaeology: Humans and Animals in Prehistory* (Cambridge, UK: Cambridge University Press, 2012), 198–199.

26 **Just like Greenland Inuits:** Pavel Nikolskiy and Vladimir Pitulko, "Evidence from the Yana Palaeolithic Site, Arctic Siberia, Yields Clues to the Riddle of Mammoth Hunting," *Journal of Archaeological Science* 40 (2013): 4189–4197.

26 **"Hunting mammoths could:** Grégory Bayle and Stéphane Péan, interview by author, Nemours, France, February 7, 2014.

26 **It's true that:** Robert Foley, "The Evolutionary Consequences of Increased Carnivory in Hominids," in *Meat-Eating and Human Evolution*, eds. Craig Stanford and Henry T. Bunn (Oxford, UK: Oxford University Press, 2001), 310.

26 **Some anthropologists argue:** John D. Speth, *Paleoanthropology and Archaeology of Big-Game Hunting* (New York: Springer, 2010).

27 **For an hour of work:** Ibid., 88–89.

27 **Hunters from one New Guinea:** Ibid., 152.

27 **The San hunters:** Ibid., 100.

27 **Even chimps are:** Ibid., 153.

27 **Scoring an elusive prize:** Ibid.

27 **Bringing an elephant:** Kristen Hawkes, "Is Meat the Hunter's Property?," in *Meat-Eating and Human Evolution*, eds. Stanford and Bunn, 228.

28 **But hunting big game:** Ibid., 230.

28 **Some scientists believe:** Ibid., 231.

28 **The Kulina practice:** Ian Gilby et al., "No Evidence of Short-Term Exchange of Meat for Sex Among Chimpanzees," *Journal of Human Evolution* 59 (2010): 44–53.

28 **Studies show that:** Russell, *Social Zooarchaeology*, 159–160.

28 **In the words:** Henry T. Bunn, "Meat Made Us Human," in *Evolution of the Human Diet*, ed. P. Ungar (Oxford, UK: Oxford University Press, 2006), 191–211.

29 **In Australia:** Norman Owen-Smith, "Contrasts in the Large Herbivore Faunas of the Southern Continents in the Late Pleistocene and the Ecological Implications for Human Origins," *Journal of Biogeography* 40 (2013): 1215–1224.

29 **By comparison, the brains:** John S. Allen, *The Omnivorous Mind: Our Evolving Relationship with Food* (Cambridge, MA: Harvard University Press, 2012), 51–52.

29 **One widely accepted:** Leslie C. Aiello and Peter Wheeler, "The Expensive-Tissue Hypothesis: The Brain and the Digestive System in Human and Primate Evolution," *Current Anthropology* 36 (1995): 199–221.

30 **A fruit-eating *Homo erectus*:** Vaclav Smil, "Eating Meat: Evolution, Patterns, and Consequences," *Population and Development Review* 28 (2002): 599–639.

30 **If as little as 10 percent:** William R. Leonard, Marcia L. Robertson, and J. Josh Snodgrass, "Energetics and the Evolution of Brain Size in Early Homo," in *Guts and Brains: An Integrative Approach to the Hominin Record*, ed. Wil Roebroeks (Leiden, Netherlands: Leiden University Press, 2007), 36.

30 **Some paleoanthropologists argue:** Alyssa N. Crittenden, "The Importance of Honey Consumption in Human Evolution, Food and Foodways: Explorations in the History and Culture of Human Nourishment," *Food and Foodways: Explorations in the History and Culture of Human Nourishment* 19 (2011): 257–273.

31 **That's 1,900 calories:** Ibid.

31 **Another food that:** Bunn, "Meat Made Us Human," 204–207.

31 **Richard Wrangham, Harvard University:** Richard Wrangham, *Catching Fire: How Cooking Made Us Human* (London: Profile Books, 2009).

31 **In Wrangham's experiments:** Rachel N. Carmody, Gil S. Weintraub, and Richard W. Wrangham, "Energetic Consequences of Thermal and Nonthermal Food Processing," *Proceedings of the National Academy of Sciences* 108 (2011): 19199–19203.

31 **Wrangham replies that:** Richard Wrangham, phone interview by author, December 1, 2013.

32 **So even though meat:** Robert Foley, "The Evolutionary Consequences of Increased Carnivory in Hominids," in *Meat-Eating and Human Evolution*, eds. Stanford and Bunn, 324.

32 **The Machiavellian intelligence:** Richard Byrne and Andrew Whiten, *Machiavellian Intelligence: Social Expertise and the Evolution of Intellect in Monkeys, Apes, and Humans* (Oxford, UK: Clarendon Press, 1988).

32 **Neither gorillas nor:** Katharine Milton, "The Critical Role Played by Animal Source Foods in Human (Homo) Evolution," *Journal of Nutrition* 133 (2003): 3886S–3892S.

32 **In the Serengeti:** Blaire Van Valkenburgh, "The Dog-Eat-Dog World of Carnivores," in *Meat-Eating and Human Evolution*, eds. Stanford and Bunn, 106.

33 **Once human hair got sparse:** Caleb E. Finch and Craig B. Stanford, "Meat-Adaptive Genes and the Evolution of Slower Aging in Humans," *Quarterly Review of Biology* 79 (2004): 28.

33 **Many scientists believe:** Meave Leakey and Lars Werdelin, "Early Pleistocene Mammals of Africa: Background to Dispersal," in *Out of Africa I: The First Hominin Colonization of Eurasia*, ed. John Fleagle (Dordrecht, Netherlands: Springer, 2010), 3.

34 **This similarity:** Ibid., 6.

34 **Each day a meat-eating animal:** Herman Pontzer, "Ecological Energetics in Early Homo," *Current Anthropology* 53 (2012): 346–358.

34 **If we hadn't eaten that meat:** Mary C. Stiner, "Carnivory, Coevolution, and the Geographic Spread of the Genus *Homo*," *Journal of Archaeological Research* 10 (2002).

34 **Analyses of nitrogen isotope:** Hervé Bocherens, "Neanderthal Dietary Habits: Review of the Isotopic Evidence," in *Evolution of Hominin Diets: Integrating Approaches to the Study of Paleolithic Subsistence*, eds. Jean-Jacques Hublin and Michael P. Richards (Dordrecht, Netherlands: Springer, 2009), 245.

35 **This would have meant:** Fred Smith and James C. M. Ahern, *The Origins of Modern Humans: Biology Reconsidered* (Hoboken, NJ: Wiley, 2013), 308.

35 **First, the meat you buy:** Frank W. Marlowe, "Hunter-Gatherers and Human Evolution," *Evolutionary Anthropology* 14 (2005): 54–67.

35 **A 3.5-ounce strip loin steak:** Louwrens C. Hoffman and Donna-Mareè Cawthorn, "What Is the Role and Contribution of Meat from Wildlife in Providing High Quality Protein for Consumption?," *Animal Frontiers* 2 (2012): 40–53.

35 **A similar-sized beefsteak:** "Many of America's Favorite Cuts Are Lean," accessed November 28, 2014, www.beefitswhatsfordinner.com/CMDocs/BIWFD/FactSheets/Many_Of_Americas_Favorite_Cuts_Are_Lean.pdf.

35 **What's more, modern meat:** Marlowe, "Hunter-Gatherers and Human Evolution."

36 **Sixty thousand years ago:** Michael P. Richards and Erik Trinkaus, "Isotopic Evidence for the Diets of European Neanderthals and Early Modern Humans," *Proceedings of the National Academy of Sciences* 106 (2009): 16034–16039.

36 **Studies of the human genome:** John Hawks et al., "Recent Acceleration of Human Adaptive Evolution," *Proceedings of the National Academy of Sciences* 104 (2007): 20753–20758.

36 **Some of us:** Gregory Cochran and Henry Harpending, *The 10,000 Year Explosion: How Civilization Accelerated Human Evolution* (New York: Basic Books, 2009), 79.

36 **Others have extra copies:** Kaixiong Ye and Zhenglong Gu, "Recent Advances in Understanding the Role of Nutrition in Human Genome Evolution," *Advances in Nutrition* 2 (2011): 486–496.

37 **For example, in some societies:** Michael J. O'Brien and Kevin N. Laland, "Genes, Culture, and Agriculture: An Example of Human Niche Construction," *Current Anthropology* 53 (2012): 438; Ye and Gu, "Recent Advances in Understanding."

37 **There is one gene:** Finch and Stanford, "Meat-Adaptive Genes."

37 **If you have the E4 allele:** Caleb E. Finch, "Evolution of the Human Lifespan and Diseases of Aging: Roles of Infection, Inflammation, and Nutrition," *Proceedings of the National Academy of Sciences* 107 (2010): 1718–1724.

37 **If two people:** Ibid.

37 **Wild chimps, for example:** Finch and Stanford, "Meat-Adaptive Genes," 19.

37 **Yet the E4 gene variant:** Alexander M. Kulminski et al., "Trade-off in the Effects of the Apolipoprotein E Polymorphism on the Ages at Onset of CVD and Cancer Influences Human Lifespan," *Aging Cell* 10 (2011): 533–541.

37 **Unfortunately, it also put them:** Mirkka Lahdenperä et al., "Fitness Benefits of Prolonged Post-Reproductive Lifespan in Women," *Nature* 428 (2004): 178–181, www.nature.com/nature/journal/v428/n6979/full/nature02367.html.

37 **The newer E3:** Peter de Knijff et al., "Lipoprotein Profile of a Greenland Inuit Population," *Arteriosclerosis and Thrombosis* 12 (1992): 1371–1379.

38 **As one writer joked:** Kerry Cue, "Try the Latest Paleo Diet and You Too Can Be Short, Stocky, Hairy and Smelly and Then You Die," *Canberra Times* (Australia), February 6, 2013.

38 **Their preserved skeletons:** Sarah Boesveld, "Cavemen's Healthy Living a Paleofantasy," *National Post*, March 16, 2013.

38 **After all, chimps:** Speth, *Paleoanthropology and Archaeology*, 151.

Chapter 3: The Good, the Bad, and the Heme Iron

41 **The late anthropologist:** Marvin Harris, *Good to Eat: Riddles of Food and Culture* (New York: Simon & Schuster, 1985), 19–22.

41 **We ate over three thousand:** Ibid., 20.

42 **In 1867 one French American:** Paul Du Chaillu, *Stories of the Gorilla Country: Narrated for Young People* (New York: Harper, 1867), 317.

42 **The Mekeo of New Guinea:** Mark Mosko, *Quadripartite Structures: Categories, Relations, and Homologies in Bush Mekeo Culture* (Cambridge, UK: Cambridge University Press, 1985), 49.

42 **In Uganda, locals have:** Merrill Kelley Bennett, *The World's Food: A Study of the Interrelations of World Populations, National Diets, and Food Potentials* (New York: Harper, 1954).

42 **"They practically *are* chimps":** Paul Breslin, interview by author, Philadelphia, PA, October 22, 2013.

43 **"These flies eat little:** Ibid.

43 **Stephen Simpson, professor:** Stephen Simpson and David Raubenheimer, *The Nature of Nutrition a Unifying Framework from Animal Adaptation to Human Obesity* (Princeton, NJ: Princeton University Press, 2012).

43 **Simpson told me once:** Stephen Simpson, phone interview by author, April 3, 2014.

44 **In 1824, Justus von Liebig:** Walter Gratzer, *Terrors of the Table: The Curious History of Nutrition* (Oxford, UK: Oxford University Press, 2005).

44 **To him, carbohydrates:** Kenneth J. Carpenter, "The History of Enthusiasm for Protein," *Journal of Nutrition* 116 (1986): 1364–1370.

44 **His "Liebig's Extract of Meat":** Justus Liebig and William Gregory, *Familiar Letters on Chemistry* (London: Taylor, Walton & Maberly, 1851), 441.

45 **Von Voit advised that:** Harold H. Mitchell, "Carl von Voit," *Journal of Nutrition* 13 (1937): 2–13.

45 **The idea that we need:** Carpenter, "The History of Enthusiasm for Protein."

45 **By 1944 the US Department:** Henry Sherman, *Principles of Nutrition and Nutritive Value of Food* (Washington, DC: US Department of Agriculture, 1944), 12.

45 **In Ghana *kwashiorkor* means:** Paul M. Insel, *Discovering Nutrition* (Burlington, MA: Jones & Bartlett Learning, 2013), 260.

45 **Soon the world fell:** Urban Jonsson, "The Rise and Fall of Paradigms in World Food and Nutrition Policy," *World Nutrition* (2010): 128–158.

45 **If your total dietary intake:** Richard Semba and Martin Bloem, *Nutrition and Health in Developing Countries* (Totowa, NJ: Humana Press, 2008), 15.

46 **It was soon discovered:** Geoffrey Webb, *Nutrition: A Health Promotion Approach* (London: Hodder Arnold, 2008), 261.

46 **"Simply put, our muscles:** Carlon Colker, "It's What's for Dinner: Beef for Maximal Muscle Gains," *Flex*, August 1, 2013, 72.

46 **"Muscle is made of protein:** Jim Stoppani, "Eat for Muscle," *Muscle & Fitness*, January 1, 2013, 92.

46 **The RDA (just like the British:** "Dietary Reference Intakes: Macronutrients," accessed November 27, 2014, www.iom.edu/Global/News%20 Announcements/~/media/C5CD2DD7840544979A549EC47E56A02B .ashx; "Nutrient Intakes," accessed November 27, 2014, www.gov.uk /government/uploads/system/uploads/attachment_data/file/265266 /familyfood-method-rni-12dec13.pdf.

47 **But most plants lack:** Kathryn Pinna Rolfes and Eleanor N. Whitney, *Normal and Clinical Nutrition* (Pacific Grove, CA: Brooks/Cole, 2011), 175–182.

47 **Another classic combination:** Nivaldo Tro, *Chemistry in Focus: A Molecular View of Our World* (Pacific Grove, CA: Brooks/Cole, 1998), 476.

48 **Lappé admitted in the:** Frances M. Lappé, *Diet for a Small Planet* (New York: Ballantine Books, 1991), 162.

48 **Their protein requirements:** Ronald Maughan, *Encyclopaedia of Sports Medicine* (Chichester, West Sussex, UK: Wiley-Blackwell, 2014), 136.

48 **If you go to the gym:** Kevin D. Tipton and Robert R. Wolfe, "Protein and Amino Acids for Athletes," *Journal of Sports Sciences* 22 (2004): 65–79.

48 **As Breslin told me:** Paul Breslin, interview by author, Philadelphia, PA, October 22, 2013.

49 **By 1919 Vilhjalmur Stefansson:** "Vilhjalmur Stefansson Was Called and Examined, May 8, 1919," accessed November 27, 2014, https://openlibrary .org/books/OL24661516M/Vihljalmur_Stefansson_was_called_and _examined_May_8_1919_i.e._1920; John D. Speth, *Paleoanthropology and Archaeology of Big-Game Hunting* (New York: Springer, 2010), 72.

49 **Even though there weren't:** Speth, *Paleoanthropology and Archaeology*, 77.

49 **you are a 110-pound woman:** "McDonald's USA Nutrition Facts for Popular Menu Items," accessed November 27, 2014, http://nutrition.mcdonalds .com/getnutrition/nutritionfacts.pdf.

49 **Your kidneys may stop:** Insel, *Discovering Nutrition*, 261.

49 **Studies show that:** Morgan E. Levine et al., "Low Protein Intake Is Associated with a Major Reduction in IGF-1, Cancer, and Overall Mortality in the 65 and Younger but Not Older Population," *Cell Metabolism* 19 (2014): 407–417.

49 **Up until the mid-nineteenth:** Alan Beardsworth and Teresa Keil, *Sociology on the Menu: An Invitation to the Study of Food and Society* (London: Routledge, 1997), 196.

50 **But more and more research:** Marta Zaraska, "Iron Deficiency, Even Mild Anemia, May Protect Against Malaria, TB and Cancer," *Washington Post*, November 24, 2014, E4.

50 **Many of the studies:** Mary H. Ward et al., "Heme Iron from Meat and Risk of Adenocarcinoma of the Esophagus and Stomach," *European Journal of Cancer Prevention* 21 (2012): 134–138.

50 **Among Tanzania's schoolchildren:** Rebecca Stoltzfus, "Epidemiology of Iron Deficiency Anemia in Zanzibari Schoolchildren: The Importance of Hookworms," *The American Journal of Clinical Nutrition* 65 (1997): 153–159.

50 **Evidence is starting:** Zaraska, "Iron Deficiency."

51 **Today, Western vegetarians:** Angela V. Saunders et al., "Iron and Vegetarian Diets," *Medical Journal of Australia Open* 1 (2012): 11–16; Madeleine J. Ball and Melinda A. Bartlett, "Dietary Intake and Iron Status of Australian Vegetarian Women," *The American Journal of Clinical Nutrition* 70 (1999): 353–358.

51 **Keeping anemia at bay:** "Iron in Your Diet," accessed November 27, 2014, www.swbh.nhs.uk/wp-content/uploads/2012/07/Iron-in-your-diet-ML3395.pdf.

51 **First, if someone has:** Saunders et al., "Iron and Vegetarian Diets."

51 **Second, low iron reserves:** Stefan Kiechl et al., "Body Iron Stores and the Risk of Carotid Atherosclerosis," *Circulation* 96 (1997): 3300–3307.

51 **Study after study shows:** Angela V. Saunders, Winston J. Craig, and Surinder K. Baines, "Zinc and Vegetarian Diets," *Medical Journal of Australia Open* 1 (2012): 17–21.

51 **The only places to get:** Suzanne Havala Hobbs, *Living Vegetarian for Dummies* (Hoboken, NJ: Wiley, 2010), 54.

52 **It's produced exclusively:** Harris, *Good to Eat*, 35–36.

52 **One study presented:** Jette F. Young et al., "Novel Aspects of Health Promoting Compounds in Meat," *Meat Science* 95 (2013): 904–911; Meltem Serdaroğlu et al., "The Power of Meat in 21st Century," ICoMST 2013 e-book of proceedings: 59th ICoMST (Izmir, Turkey: Ege University, 2013).

52 **Because of all the longitudinal:** Gary E. Fraser and David J. Shavlik, "Ten Years of Life," *Archives of Internal Medicine* 161 (2001): 1645–1652; Paul N. Appleby et al., "The Oxford Vegetarian Study: An Overview," *The American Journal of Clinical Nutrition* 70 (1999): 525s–531s.

52 **By January 1917:** Ina Zweiniger-Bargielowska, Rachel Duffett, and Alain Drouard, eds., *Food and War in Twentieth Century Europe* (Farnham, UK: Ashgate, 2011), 202–210; Mikkel Hindhede, "The Effect of Food Restriction During War on Mortality in Copenhagen," *JAMA* 74 (1920): 381–382.

53 **The vegetarian Seventh-day Adventists:** Fraser and Shavlik, "Ten Years of Life."

53 **A fruitarian diet can be:** "Top Diets Review for 2014," accessed November 27, 2014, www.nhs.uk/Livewell/loseweight/Pages/top-10-most-popular-diets-review.aspx#dukan.

54 **According to this theory:** Iztok Ostan, "Appetite for the Selfish Gene," *Appetite* 54 (2010): 442–449; Imogen S. Rogers et al., "Diet Throughout Childhood and Age at Menarche in a Contemporary Cohort of British Girls," *Public Health Nutrition* 13 (2010): 2052–2063.

Chapter 4: The Chemistry of Love: Umami, Aromas, and Fat

58 **When prisoners on death row:** Brian Wansink, Kevin M. Kniffin, and Mitsuru Shimizu, "Death Row Nutrition: Curious Conclusions of Last Meals," *Appetite* 59 (2012): 837–843.

58 **Studies show that 74 percent:** Jamie L. Osman and Jeffery Sobal, "Chocolate Cravings in American and Spanish Individuals: Biological and Cultural Influences," *Appetite* 47 (2006): 290–301.

58 **"Oh, yes, cheesesteaks:** Gary Beauchamp, interview by author, Philadelphia, PA, October 2, 2013.

58 **When I ask him what:** Ibid.

59 **What we do know:** Danielle R. Reed, Toshiko Tanaka, and Amanda H. McDaniel, "Diverse Tastes: Genetics of Sweet and Bitter Perception," *Physiology & Behavior* 88 (2006): 215–226.

59 **It happens like this:** Barb Stuckey, *Taste: Surprising Stories and Science About Why Food Tastes Good* (New York: Simon & Schuster, 2012), 47–48.

60 **The discovery of why:** Jozef Cohen, "Taste Blindness to Phenyl-thiocarbamide and Related Compounds," *Psychological Bulletin* 46 (1949): 490–498; Linda Bartoshuk, "Comparing Sensory Experiences Across Individuals: Recent Psychophysical Advances Illuminate Genetic Variation in Taste Perception," *Chemical Senses* 25 (2000): 447–460.

60 **"He had lung cancer:** Linda Bartoshuk, phone interview by author, August 27, 2013.

60 **She started noticing:** Ibid.

61 **Nontasters, like myself:** Adam Drewnowski, Susan Ahlstrom Henderson, and Anne Barratt-Fornell, "Genetic Taste Markers and Food Preferences," *Drug Metabolism and Disposition* 29 (2001): 535–538.

61 **They also tend to eat:** Drewnowski, Henderson, and Barratt-Fornell, "Genetic Taste Markers"; Danielle Renee Reed and Antti Knaapila, "Genetics of Taste and Smell: Poisons and Pleasures," *Progress in Molecular Biology and Translational Science* 94 (2010): 213–240.

62 **Some studies do show:** Bianca Turnbull and Elizabeth Matisoo-Smith, "Taste Sensitivity to 6-n-propylthiouracil Predicts Acceptance of Bitter-Tasting Spinach in 3–6-y-old Children," *The American Journal of Clinical Nutrition* 76 (2002): 1101–1105; Blanca J. Villarino et al., "Relationship of PROP (6-N-Propylthiouracil) Taster Status with the Body Mass Index and Food Preferences of Filipino Adults," *Journal of Sensory Studies* 24 (2009): 354–371.

62 **According to Bartoshuk:** Bartoshuk, interview.

62 **A large chunk of what:** Stuckey, *Taste*, 54–55.

62 **The lack of strong aromas:** Leo Nollet and Fidel Toldra, eds., *Handbook of Muscle Foods Analysis* (Boca Raton, FL: CRC Press, 2009), 504–505.

62 **Even animals seem:** Rachel N. Carmody, Gil S. Weintraub, and Richard W. Wrangham, "Energetic Consequences of Thermal and Nonthermal Food Processing," *Proceedings of the National Academy of Sciences* 108 (2011): 19199–19203 (see Kristen Hawkes, "Is Meat the Hunter's Property?," in *Meat-Eating and Human Evolution*, eds. Stanford and Bunn, 231.)

62 **In similar experiments:** Victoria Wobber, Brian Hare, and Richard Wrangham, "Great Apes Prefer Cooked Food," *Journal of Human Evolution* 55 (2008): 1–9.

62 **Although Maillard's name is now:** "Who Is Louis Camille Maillard?," accessed November 28, 2014, www.lc-maillard.org/who-is-louis-camille-maillard/.

63 **Only later did he begin:** Tom Jaine, ed., *Oxford Symposium on Food & Cookery 1987: Taste* (London: Prospect, 1988), 134.

63 **There are over one thousand:** Nollet and Toldra, *Handbook*, 504.

63 **Some smell fruity:** Virginia C. Resconi, Ana Escudero, and María M. Campo, "The Development of Aromas in Ruminant Meat," *Molecules* 18 (2013): 6748–6781; Chris Calkins and Jennie Marie Hodgen, "A Fresh Look at Meat Flavor," *Meat Science* 77 (2007): 63–80; "Odors and Odorants of the Maillard Class," accessed November 28, 2014, www.flavornet.org/class/maillard.html.

63 **There are several passages:** Paul Breslin, interview by author, Philadelphia, PA, October 22, 2013.

63 **One explanation is that:** Ibid.

63 **Other products of the Maillard:** Richard H. Stadler et al., "Food Chemistry: Acrylamide from Maillard Reaction Products," *Nature* 419 (2002): 449–450; Frederic Tessier and Inès Birlouez-Aragon, "Health Effects of Dietary Maillard Reaction Products: The Results of ICARE and Other Studies," *Amino Acids* 42 (2012): 1119–1131.

64 **It is more energy dense:** John Prescott, *Taste Matters: Why We Like the Foods We Do* (London: Reaktion Books, 2012), 37.

64 **"When you think you're:** Breslin, interview.

64 **Fat is also where:** David B. Min, Thomas H. Smouse, and Stephen S. Chang, eds., *Flavor Chemistry of Lipid Foods* (Champaign, IL: American Oil Chemists' Society, 1989), 172.

64 **Boiled or stewed beef smells:** "The Chemistry of Beef Flavor," accessed November 28, 2014, http://beefresearch.org/CMDocs/BeefResearch/The%20Chemistry%20of%20Beef%20Flavor.pdf.

64 **Another potent compound:** Donald Mottram, "Flavour Formation in Meat and Meat Products: A Review," *Food Chemistry* 62 (1998): 415–424; Gengjun Chen, Huanlu Song, and Changwei Ma, "Aroma-Active Compounds of Beijing Roast Duck," *Flavour and Fragrance Journal* 24 (2009): 186–191.

64 **Chicken, meanwhile, smells:** Leo Nollet, *Handbook of Meat, Poultry and Seafood Quality* (Ames, IA: Blackwell, 2007), 140; "Trans, trans-2, 4-decadienal," accessed November 28, 2014, www.ymdb.ca/compounds/YMDB01455.

64 **Neural-imaging studies show:** Beverly J. Tepper, "The Taste for Fat: New Discoveries on the Role of Fat in Sensory Perception, Metabolism, Sensory Pleasure, and Beyond," *Journal of Food Science* 77 (2012): vi.

64 **What's more, over the last:** Ibid.

65 **The more mushroom-like:** René Nachtsheim and Elmar Schlich, "The Influence of 6-n-propylthiouracil Bitterness, Fungiform Papilla Count and Saliva Flow on the Perception of Pressure and Fat," *Food Quality and Preference* 29 (2013): 137–145.

65 **But nontasters have:** René Nachtsheim and Elmar Schlich, "The Influence of Oral Phenotypic Markers and Fat Perception on Fat Intake During a Breakfast Buffet and in a 4-Day Food Record," *Food Quality and Preference* 32 (2014): 173–183.

65 **Because of its very active:** Colin Schultz, "What Did Dinosaur Taste Like?," *Smithsonian*, December 28, 2012, accessed January 17, 2015, www.smithsonianmag.com/smart-news/what-did-dinosaur-taste-like-28428/?no-ist.

66 **In Kyoto, where Ikeda:** "Kikunae Ikeda (Discoverer of 'Umami')," accessed December 1, 2014, www.s.u-tokyo.ac.jp/en/research/alumni/ikeda.html.

66 **He quickly obtained a patent:** Ibid.

67 **It was only in 2000:** "Umami Taste Receptor Identified," February 2000, accessed December 1, 2014, www.nature.com/neuro/press_release/nn0200.html.

67 **When one or both:** "What Is Umami?," Umami Information Center, accessed December 1, 2014, www.umamiinfo.com/2013/02/tasting-umami-c2.php.

67 **Meat is a particularly good:** Marta Zaraska, "What Makes a Hamburger and Other Cooked Meat So Enticing to Humans?," *Washington Post*, August 13, 2013, E4.

67 **All this is no secret:** Jacqueline B. Marcus, "Unleashing the Power of Umami," *Food Technology* 63 (2009).

67 **It maximizes the delicious:** Stuckey, *Taste*.

67 **We learn its taste:** Stuckey, *Taste*, 247; Gary Beauchamp, "Sensory and Receptor Responses to Umami: An Overview of Pioneering Work," *The American Journal of Clinical Nutrition* 90 (2009): 723S–727S.

68 **If the soup experiment doesn't:** Olivia Lugaz, Anne-Maria Pillias, and Annick Faurion, "A New Specific Ageusia: Some Humans Cannot Taste L-glutamate," *Chemical Senses* 27 (2002): 105–115.

68 **Meanwhile, a study of twins:** Fiona M. Breena, Robert Plomin, and Jane Wardle, "Heritability of Food Preferences in Young Children," *Physiology & Behavior* 88 (2006): 443–447.

68 **Other experiments point:** Jane Wardle and Lucy J. Cooke, "One Man's Meat Is Another Man's Poison," *EMBO Reports* 11 (2010): 816–821; Danielle Renee Reed and Antti Knaapila, "Genetics of Taste and Smell: Poisons and Pleasures," *Progress in Molecular Biology and Translational Science* 94 (2010): 213–240.

69 **The thing about:** Hongyu Zhao et al., "Pseudogenization of the Umami Taste Receptor Gene Tas1r1 in the Giant Panda Coincided with Its Dietary Switch to Bamboo," *Molecular Biology and Evolution* 27 (2010): 2669–2673.

69 **"What we do know:** Beauchamp, interview.

70 **According to one survey:** Hank Rothgerber, "Efforts to Overcome Vegetarian-Induced Dissonance Among Meat Eaters," *Appetite* 79 (2014): 32–41.

CHAPTER 5: WHY WOULD ABRAMOVICH TASTE GOOD?

73 **"Come and get your:** Ifor Humphreys, interview by author, Abermule, UK, July 15, 2014.

74 **According to a legend:** "The History of the Berkshire Breed," accessed December 1, 2014, www.ansi.okstate.edu/breeds/swine/.

74 **The official website:** Kagoshima Prefecture, www.pref.kagoshima.jp.

74 **The tenderness of meat:** Elton Aberle et al., *Principles of Meat Science* (Dubuque, IA: Kendall Hunt, 2012), 261.

75 **On the other hand:** Ibid.

75 **Not everyone agrees:** Ibid., 79.

75 **In the West, though:** Angela Reicks et al., "Demographics and Beef Preferences Affect Consumer Motivation for Purchasing Fresh Beef Steaks and Roasts," *Meat Science* 87 (2011): 403–411.

75 **To be tender:** Aberle et al., *Principles of Meat Science*, 261.

75 **Such fat gets released:** Ibid., 263.

75 **When Mark Schatzker:** Mark Schatzker, *Steak: One Man's Search for the World's Tastiest Piece of Beef* (New York: Viking, 2010), 159.

76 **"In all of UK there is:** Humphreys, interview.

76 **Seven years ago:** Ibid.

76 **Only a chosen few:** "Classification," Kobe Beef Marketing & Distribution Promotion Association, accessed December 1, 2014, www.kobe-niku.jp/en/contents/about/criteria.html.

76 **Either a pure-blood bull:** Ibid.

76 **Each year as few as:** "FAQ," Kobe Beef Marketing & Distribution Promotion Association, accessed December 1, 2014, www.kobe-niku.jp/en/contents/faq/index.html.

76 **The first cut of true Kobe:** "Exported Beef," Kobe Beef Marketing & Distribution Promotion Association, accessed December 1, 2014, www.kobe-niku.jp/en/contents/exported/index.php?y=2012; Michael Shallberg, Fremont Beef Company, e-mail message to author, February 13, 2014.

76 **In all of 2013, just:** "Exported Beef."

77 **Since cattle of the Angus-Aberdeen:** Aberle, *Principles of Meat Science*, 85.

77 **It has to have:** "Our 10 Quality Specifications," Certified Angus Beef LLC, accessed December 2, 2014, www.certifiedangusbeef.com/brand/specs.php.

77 **The argument behind:** Rhonda Miller, "Factors Affecting the Quality of Raw Meat," in *Meat Processing: Improving Quality*, eds. Joseph Kerry, John Kerry, and David Ledward (Cambridge, UK: Woodhead Publishing, 2002), 43.

77 **In those days:** Aberle, *Principles of Meat Science*, 91.

77 **A majority of Americans:** Bethany Sitz et al., "Consumer Sensory Acceptance and Value of Domestic, Canadian, and Australian Grass-Fed Beef Steaks," *Journal of Animal Science* 83 (2005): 2863–2868; Schatzker, *Steak*, 41.

78 **One problem American consumers:** Paul Warriss, *Meat Science: An Introductory Text* (Wallingford, UK: CABI, 2010), 79.

78 **According to the official:** "FAQ."

78 **"It's a good story:** Humphreys, interview.

78 **If you find that your sautéed pork:** Warriss, *Meat Science*, 104–109.

79 **Once an aspiring vet:** Edward Mills, interview by author, State College, PA, October 10, 2013.

80 **Intact boars don't get:** Warriss, *Meat Science*, 81.

80 **If boars could choose:** Ibid., 151.

80 **That the animals suffer:** "Bacon: A Day in the Life," Free from Harm, December 18, 2013, accessed December 2, 2014, http://freefromharm.org/animal-cruelty-investigation/day-in-the-life-christmas-ham-pig/.

80 **Its electrodes are applied:** Warriss, *Meat Science*, 54–55.

80 **Every Tuesday between:** Dawne McCance, *Critical Animal Studies: An Introduction* (Albany: State University of New York Press, 2013), 18; "Slaughterhouse with Capacity 46,000 Birds/Hour," SPEKTR, accessed December 2, 2014, www.spektr.kiev.ua/en/index.php/poultry-and-livestock-complexes/slaughterhouse-with-capacity-46-000-birds-hour.

80 **Every day in the US:** "Farm Animal Statistics: Slaughter Totals," The Humane Society, accessed December 3, 2014, www.humanesociety.org/news/resources/research/stats_slaughter_totals.html.

81 **"If the animal is severely:** Mills, interview.

81 **In its final moment:** Aberle, *Principles of Meat Science*, 83.

81 **In the US, 16 percent:** Warriss, *Meat Science*, 104–105.

81 **Once "they die piece by piece":** Joby Warrick, "'They Die Piece by Piece': In Overtaxed Plants, Humane Treatment of Cattle Is Often a Battle Lost," *Washington Post*, April 10, 2001, A01; Aberle et al., *Principles of Meat Science*, 108.

81 **Making matters worse:** Ibid., 105–107.

82 **It may work to make:** Warriss, *Meat Science*, 79.

82 **Some of its muscles:** Ibid., 107.

83 **A few years ago:** Mills, interview.

83 **Cold shortening happens:** Warriss, *Meat Science*, 115.

83 **Imagine taking a bundle:** Mills, interview.

83 **Sometimes meat producers:** Ibid.

84 **To produce pale veal:** Warriss, *Meat Science*, 153.

84 **Such calves can't move:** Report on the Welfare of Calves, 1995, European Commission, accessed May 11, 2015, http://ec.europa.eu/food/fs/sc/oldcomm4/out35_en.pdf.

84 **The American Veal Association:** "Veal FAQ," The American Veal Association, accessed December 2, 2014, www.americanveal.com/for-consumers/veal-frequently-asked-questions/.

84 **Now one of her pet projects:** Temple Grandin, phone interview by author, February 26, 2013.

85 **Add some Optaflexx:** "Optaflexx," Elanco, 2012, accessed December 3, 2014, www.elanco.us/products-services/beef/improve-feedlot-cattle-weight-gain-efficiency.aspx.

85 **Meat from animals fed:** Warriss, *Meat Science*, 20.

85 **Seventy percent of US:** Tom Polansek and P. J. Huffstutter, "Halt in Zilmax Sales Fuels Demand for Rival Cattle Feed Product," Reuters, August 23, 2013, accessed December 3, 2014, www.reuters.com/article/2013/08/23/livestock-zilmax-lilly-idUSL2N0GO10U20130823.

85 **"Hot weather makes:** Grandin, interview.

85 **Such cows, whose feet:** P. J. Huffstutter and Tom Polansek, "Special Report: Lost Hooves, Dead Cattle Before Merck Halted Zilmax Sales," Reuters, December 31, 2013, accessed July 24, 2015, www.reuters.com/article/2013/12/31/us-zilmax-merck-cattle-special-report-idUSBRE9BT0NV20131231.

85 **To achieve this:** "The Chemistry of Beef Flavor," The National Cattlemen's Beef Association, accessed December 3, 2014, http://beefresearch.org/CMDocs/BeefResearch/The%20Chemistry%20of%20Beef%20Flavor.pdf.

85 **They not only make:** Ibid.

86 **If injecting the meat:** Peter Sheard, "Processing and Quality Control of Restructured Meat," in *Meat Processing*, eds. Kerry, Kerry, and Ledward, 332–353.

86 **(If you want to trick:** Ibid.

86 **You are most likely:** Sheard, "Processing and Quality Control," 332–353; Jim Hightower, "What's Really in Your Steak?," Salon, June 8, 2012, accessed December 3, 2014, www.salon.com/2012/06/08/whats_really_in_your_steak_salpart/.

86 **A survey in the UK:** Sheard, "Processing and Quality Control," 332.

86 **They research feeding:** Warriss, *Meat Science*, 55.

87 **Although the taste:** Eleanor Mackay, "Price Most Important Factor for Meat Buying Public," July 21, 2014, accessed December 3, 2014, http://meatinfo.co.uk/news/fullstory.php/aid/17205/Price_most_important_factor_for_meat_buying_public.html.

CHAPTER 6: WAGGING THE DOG OF DEMAND

89 **Some researchers argue:** Marta G. Rivera-Ferre, "Supply vs. Demand of Agri-industrial Meat and Fish Products: A Chicken and Egg Paradigm?," *International Journal of Sociology of Agriculture and Food* 16 (2009): 90–105.

90 **The industry doesn't exactly:** Amanda Radke, "New Beef, Pork Names Don't Do Us Any Special Favors," *Beef*, June 27, 2013.

90 **Consider these numbers:** "U.S. Meat and Poultry Production & Consumption: An Overview," American Meat Institute, January 2015, accessed August 24, 2015, www.meatami.com/ht/a/GetDocumentAction/i/93335.

90 **However, just four pork producers:** Shefali Sharma, "Food and National Security: The Shuanghui-Smithfield Merger Revisited," Institute for Agriculture and Trade Policy, September 12, 2013, accessed December 2, 2014, www.iatp.org/blog/201309/food-and-national-security-the-shuanghui-smithfield-merger-revisited#sthash.oGoFpEkt.JqIBctu5.dpuf; Roberto Ferdman, "Americans Have Never Had So Few Options in Deciding What Company Makes Their Meat," *Washington Post*, September 4, 2014, accessed December 3, 2014, www.washingtonpost.com/blogs/wonkblog/wp/2014/06/11/americans-have-never-had-so-few-options-in-deciding-what-company-makes-their-meat/.

90 **Tyson, the largest meat corporation:** David Kesmodel and Laurie Burkitt, "Inside China's Supersanitary Chicken Farms," *Wall Street Journal*, December 9, 2013, accessed December 3, 2014, http://online.wsj.com/articles/SB10001424052702303559504579197662165181956.

90 **According to the American:** "The United States Meat Industry at a Glance," American Meat Institute, accessed December 2, 2014, www.meatami.com/ht/d/sp/i/47465/pid/47465.

91 **Even if the fruit and vegetable:** "Cash Receipts for Corn and Soybeans Account for About Half of All Crop Receipts," USDA, September 16,

2013, accessed December 3, 2014, www.ers.usda.gov/data-products/chart-gallery/detail.aspx?chartId=40050&ref=collection&embed=True#.VCEp3vZ8Gxo.

91 **That's almost four times:** "Farming and Farm Income," USDA, September 9, 2014, accessed December 3, 2014, www.ers.usda.gov/data-products/ag-and-food-statistics-charting-the-essentials/farming-and-farm-income.aspx#.VCEqbPZ8Gxo.

91 **Beans, peas, and lentils:** "Vegetables and Pulses Yearbook Data," USDA, May 30, 2014, accessed December 3, 2014, www.ers.usda.gov/datafiles/Vegetable_and_Pulses_Yearbook_Tables/General/YRBK2014_Section%201_General.pdf.

91 **In the US each beef producer:** "Beef Checkoff Questions and Answers," Cattlemen's Beef Board, accessed December 3, 2014, www.beefboard.org/about/faq_aboutwhopays.asp; "About Pork Checkoff and the National Pork Board," National Pork Board, accessed December 3, 2014, www.pork.org/about-us/#.VCQXJfZ8Gxo.

91 **In Canada, the levy is $1:** Ibid.

91 **To give you some perspective:** Marion Nestle, *Food Politics: How the Food Industry Influences Nutrition and Health* (Berkeley: University of California Press, 2002), 131.

91 **Back in 1992:** "Contemporary Beef Marketing Campaign Builds on Popular, Successful Tagline," Cattlemen's Beef Board, June 23, 2014, accessed December 3, 2014, www.beefboard.org/news/140623BIWFDHistoryOfFeatureStory.asp.

91 **As for its effectiveness:** Ibid.

91 **In 2015 the beef industry:** "Beef Checkoff Sets FY 2015 Plan of Work," Cattlemen's Beef Board, accessed December 3, 2014, www.beefboard.org/news/140918OCSetsFY15PlanofWork.asp.

92 **Yet it's the youngest consumers:** "Authorization Request for FY 2015," Cattlemen's Beef Board, accessed December 3, 2014, www.beefboard.org/member-toolkit/files/Operating%20Committee/September%202014%20OC%20Meeting/1507-CI%20ANCW.pdf.

92 **To encourage them to eat:** "Work Plan for FY 2014," Cattlemen's Beef Board, accessed December 3, 2014, www.beefboard.org/blog/2013 Summer Meeting Materials/Work Plans/ANCW Work Plan 2014 Updated2.docx.

92 **They create apps:** "Tips to Market Beef to Millennials," Cattlemen's Beef Board, accessed December 3, 2014, www.beefretail.org/themillennialshopper.aspx.

92 **Between 2006 and 2013:** "Beef Checkoff Shows Strong Returns on the Dollar," Farm and Dairy, August 5, 2014, accessed December 3, 2014, www

.farmanddairy.com/news/beef-checkoff-shows-strong-returns-dollar/205826.html.

92 **If it wasn't for the checkoff:** Ibid.

92 **Meanwhile, the pork industry's:** "The Other White Meat® Brand," National Pork Board, accessed December 3, 2014, www.porkbeinspired.com/about-the-national-pork-board/the-other-white-meat-brand/.

92 **During the five years:** Trish Hall, "And This Little Piggy Is Now on the Menu," *New York Times*, November 13, 1991.

92 **As C. W. Post:** Paul Fieldhouse, *Food and Nutrition: Customs and Culture* (Cheltenham, UK: Nelson Thornes, 1998), 11.

92 **As David Robinson Simon writes:** David R. Simon, *Meatonomics: How the Rigged Economics of Meat and Dairy Make You Consume Too Much and How to Eat Better, Live Longer, and Spend Smarter* (San Francisco: Conari Press, 2013), 6.

93 **As Bill Roenigk explained:** Bill Roenigk, interview by author, Washington, DC, October 28, 2013.

93 **Although it's not a meat:** Joe Roybal, "Big Beef Buyers," *Beef*, February 1, 2007, accessed December 3, 2014, http://beefmagazine.com/mag/beef_big_beef_buyers; "Irish Beef," McDonald's Corporation, accessed December 3, 2014, www.mcdonalds.ie/iehome/food/food_quality/about_our_food/our_beef.html.

93 **Selling on average:** Amanda Radke, "McDonald's Wants Industry Help in Defining Sustainable Beef," Beef Daily, August 20, 2014, accessed December 3, 2014, http://beefmagazine.com/blog/mcdonald-s-wants-industry-help-defining-sustainable-beef.

93 **In 2011 it spent:** Christina Austin, "The Billionaire's Club: Only 36 Companies Have $1,000 Million–Plus Ad Budgets," Business Insider, November 11, 2012, accessed December 3, 2014, www.businessinsider.com/the-35-companies-that-spent-1-billion-on-ads-in-2011-2012-11?op=1#ixzz3EPZEKhQW.

93 **And guess which ads:** Ameena Batada et al., "Nine out of 10 Food Advertisements Shown During Saturday Morning Children's Television Programming Are for Foods High in Fat, Sodium, or Added Sugars, or Low in Nutrients," *Journal of the American Dietetic Association* 108 (2008): 673–678.

93 **The only figure that American:** Andrew Smith, *Hamburger: A Global History* (London: Reaktion, 2008), 53.

93 **"Rather to make the consumer:** Elin Kubberød et al., "The Effect of Animality on Disgust Response at the Prospect of Meat Preparation—An Experimental Approach from Norway," *Food Quality and Preference* 17 (2006): 199–208.

94 **Consultants, lawyers, and:** Jeffrey H. Birnbaum, "The Road to Riches Is Called K Street," *Washington Post*, June 22, 2005, A01.

94 **The Center for Responsive Politics:** "Lobbying Database," Center for Responsive Politics, accessed December 3, 2014, www.opensecrets.org/lobby/alphalist_indus.php.

94 **And such contributions:** Thomas Stratmann, "Can Special Interests Buy Congressional Votes? Evidence from Financial Services Legislation," *Journal of Law and Economics* 45 (2002): 345–373.

94 **According to Chuck Conner:** "Hearing to Review the Proposal of the United States Department of Agriculture for the 2007 Farm Bill with Respect to Specialty Crops and Organic Agriculture," US Government Printing Office, accessed December 4, 2014, www.gpo.gov/fdsys/pkg/CHRG-110hhrg48113/html/CHRG-110hhrg48113.htm.

94 **Meanwhile, between 1995:** "Livestock Subsidies," EWG Farm Subsidies, accessed December 4, 2014, http://farm.ewg.org/progdetail.php?fips=00000&progcode=livestock.

94 **The author of *Meatonomics*:** Simon, *Meatonomics*, xv.

94 **From 1995 to 2012:** "Corn Subsidies," EWG Farm Subsidies, accessed December 4, 2014, http://farm.ewg.org/progdetail.php?fips=00000&progcode=corn; "Soybean Subsidies," EWG Farm Subsidies, accessed December 4, 2014, http://farm.ewg.org/progdetail.php?fips=00000&progcode=soybean.

94 **Since 60 percent:** "Below-Cost Feed Crops," Institute for Agriculture and Trade Policy, accessed December 4, 2014, www.iatp.org/files/258_2_88122_0.pdf.

95 **"NCC did a study:** Roenigk, interview.

95 **And then the demand:** Tatiana Andreyeva, Michael Long, and Kelly Brownell, "The Impact of Food Prices on Consumption: A Systematic Review of Research on the Price Elasticity of Demand for Food," *The American Journal of Public Health* 100 (2010): 216–222; Harry Kaiser, "An Economic Analysis of the National Pork Board Checkoff Program," April 12, 2012, accessed December 4, 2014, www.pork.org/filelibrary/ReturnOnInvestment/2011ROIStudyCornellUnivDrHarryKaiserReportComplete.pdf.

95 **However, some consumers:** Thom Blischok, "Creating a Successful Consumer Market Strategy," National Chicken Council, 2011, accessed December 4, 2014, www.nationalchickencouncil.org/wp-content/uploads/2012/01/Blischok-Symphony-IRI-NCC-Annual-Conference-2011.pdf.

95 **Neal Barnard, professor:** Neal Barnard, Andrew Nicholson, and J. L. Howard, "The Medical Costs Attributable to Meat Consumption," *Preventive Medicine* 24 (1995): 646–655.

95 **In *Meatonomics*, Simon:** Simon, *Meatonomics*, xx.

95 **For every dollar of beef:** Ibid.

95 **So the next time you:** Jules Pretty et al., "Farm Costs and Food Miles: An Assessment of the Full Cost of the UK Weekly Food Basket," *Food Policy* 30 (2005): 1–19.

96 **On June 3, 2013:** David Rogers, "Bye-bye 'Meatless Mondays,'" POLITICO, June 25, 2013, accessed December 4, 2014, www.politico.com/story/2013/06/bye-bye-meatless-mondays-93349.html#ixzz3E9GZnxkS.

96 **As Marion Nestle, professor:** Marion Nestle, e-mail message to author, June 20, 2014.

96 **The guidelines, according:** www.choosemyplate.gov/dietary-guidelines.html.

96 **In her book *Food Politics*:** Nestle, *Food Politics*, 30.

96 **Over the years the standard:** Ibid., 3.

97 **Over the years, some:** Jeff Herman, "Saving U.S. Dietary Advice from Conflicts of Interest," *Food and Drug Law Journal* 65 (2010): 309–316.

97 **Dale Moore, chief of staff:** Philip Mattera, "USDA INC.: How Agribusiness Has Hijacked Regulatory Policy at the U.S. Department of Agriculture," July 23, 2004, accessed December 4, 2014, www.competitivemarkets.com/wp-content/uploads/2004/07/USDAINC_AG_HIJACKING.pdf.

98 **For example, the author:** Shalene McNeilla and Mary Van Elswyk, "Red Meat in Global Nutrition," *Meat Science* 92 (2012): 166–173.

98 **The author of yet another:** Mike Roussell et al., "Effects of a DASH-like Diet Containing Lean Beef on Vascular Health," *Journal of Human Hypertension* 28 (2014): 600–605.

98 **A Swedish study:** Christel Larsson and Gunnar Johansson, "Dietary Intake and Nutritional Status of Young Vegans and Omnivores in Sweden," *American Journal of Clinical Nutrition* 76 (2002): 100–106.

98 **Tyson Foods, the California:** American Heart Association, "2013 American Stroke Association's Annual Report," accessed December 4, 2014, www.heart.org/HEARTORG/General/2013-2014-Annual-Report_UCM_448427_Article.jsp.

98 **The American Dietetic Association**: "American Dietetic Association/ADA Foundation 2010 Annual Report," American Dietetic Association, 2010, accessed August 24, 2015, http://www.eatrightpro.org/~/media/eatrightpro%20files/about%20us/annual%20reports/2010_ada_annual_report.ashx; "2013 Academy of Nutrition and Dietetics Foundation Donor Report," accessed December 4, 2014, www.eatright.org/workarea/linkit.aspx?linkidentifier=id&itemid=6442473017&libid=6442472995; www.eatright.org/WorkArea/linkit.aspx?LinkIdentifier=id&ItemID=6442473017&libID=6442472995.

98 **The editors wrote that:** "Journal Policy on Research Funded by the Tobacco Industry," *BMJ*, October 15, 2013, accessed December 4, 2014, www.bmj.com/press-releases/2013/10/15/bmj-journal-editors-will-no-longer-consider-research-funded-tobacco-indust.

98 **Research done on:** Simon, *Meatonomics*, 11.

99 **According to the pork:** Kaiser, "An Economic Analysis."

99 **This may seem like:** "Pork Facts," National Pork Producers Council, accessed December 4, 2014, www.nppc.org/pork-facts/.

99 **As he told me:** T. Colin Campbell, phone interview by author, July 8, 2014.

100 **At some point:** Sam Howe Verhovek, "Talk of the Town: Burgers v. Oprah," *New York Times*, January 21, 1998, 10.

100 **"Burgers v. Oprah":** Ibid.

100 **In thirteen states:** Rita Marie Cain, "Food, Inglorious Food: Food Safety, Food Libel, and Free Speech," *American Business Law Journal* 49 (2012): 275–324.

100 **Although Winfrey and:** Nestle, *Food Politics*, 164.

100 **As Robert Hatherill, author:** J. Robert Hatherill, "Take the Gag Off Food Safety Issues," *Los Angeles Times*, April 12, 1999.

100 **When I talk to Lyman:** Howard Lyman, phone interview by author, June 2, 2014.

101 **The other ones are:** Mark Bittman, "Who Protects the Animals?," Opinionator blog, *New York Times*, April 26, 2011.

101 **In the past, investigations:** Ibid.; Michelle Kretzer, "Video Shows Pigs Mutilated, Beaten, Duct-Taped," PETA, October 12, 2010, accessed December 4, 2014, www.peta.org/blog/pigs-mutilated-beaten-duct-taped/; "KFC's Cut-Throat Killing Machine," PETA, accessed December 4, 2014, www.peta.org/blog/kfc-s-cut-throat-killing-machine/.

101 **In Iowa, for example:** Cody Carlson, "The Ag Gag Laws: Hiding Factory Farm Abuses from Public Scrutiny," *The Atlantic*, March 20, 2012, accessed December 4, 2014, www.theatlantic.com/health/archive/2012/03/the-ag-gag-laws-hiding-factory-farm-abuses-from-public-scrutiny/254674/.

101 **As the meat magnate:** Maureen Ogle, *In Meat We Trust: An Unexpected History of Carnivore America* (Boston: Houghton Mifflin Harcourt, 2013), 60.

Chapter 7: Eating Symbols

104 **In one large study:** Anne Lanfer et al., "Predictors and Correlates of Taste Preferences in European Children: The IDEFICS Study," *Food Quality and Preference* 27 (2013): 128–136.

104 **Studies show, for instance:** Julie Mennella, Coren Jagnow, and Gary Beauchamp, "Prenatal and Postnatal Flavor Learning by Human Infants," *Pediatrics* 107 (2001): e88.

104 **If you don't care:** Benoist Schaal, Luc Marlier, and Robert Soussignan, "Human Foetuses Learn Odours from Their Pregnant Mother's Diet," *Chemical Senses* 25 (2000): 729–737.

104 **As a result, rats develop:** Robert Boyd and Peter Richerson, *The Origin and Evolution of Cultures* (Oxford, UK: Oxford University Press, 2005), 425–426.

104 **But rats and humans:** Ibid.

105 **In experiments, kittens:** Leann Birch, "Development of Food Preferences," *Annual Review of Nutrition* 19 (1999): 41–62.

105 **Studies show that:** Laetitia Barthomeuf, Sylvie Droit-Volet, and Sylvie Rousset, "How Emotions Expressed by Adults' Faces Affect the Desire to Eat Liked and Disliked Foods in Children Compared to Adults," *British Journal of Developmental Psychology* 30 (2012): 253–266.

105 **Brian Wansink, an expert:** Brian Wansink, phone interview by author, September 30, 2014.

106 **For this reason, meat:** Paul Fieldhouse, *Food and Nutrition: Customs and Culture* (Cheltenham, UK: Nelson Thornes, 1998), 93.

107 **"Blood and meat eating is:** Paul Akakpo, interview by author, Ouidah, Benin, March 27, 2010; Paul Akakpo, e-mail message to author, February 23, 2015; Judy Rosenthal, *Possession, Ecstasy, and Law in Ewe Voodoo* (Charlottesville: University Press of Virginia, 1998), 236.

107 **In one of his:** Carol Nemeroff and Paul Rozin, "'You Are What You Eat': Applying the Demand-Free 'Impressions' Technique to an Unacknowledged Belief," *Ethos* 17 (1989): 50–69.

108 **If you eat a lot:** Ibid.

108 **You may have seen:** "Tofu (Hummer Ad)," July 27, 2006, accessed December 5, 2014, www.youtube.com/watch?v=lL4ZkYPLN38.

108 **What I've just described:** Carol J. Adams, *Sexual Politics of Meat: A Feminist-Vegetarian Critical Theory* (New York: Continuum, 2010), 5.

108 **Domino's Pizza did one:** Richard Rogers, "Beasts, Burgers, and Hummers: Meat and the Crisis of Masculinity in Contemporary Television Advertisements," *Environmental Communication* 2 (2008): 281–301.

109 **In Burger King's commercial:** "Burger King Manthem Commercial," July 26, 2007, accessed December 5, 2014, www.youtube.com/watch?v=R3YHrf9fGrw.

109 **In New Zealand, a campaign:** "The New Currency of Man-Lion Red's Man Points," Scoop, February 10, 2011, accessed December 5, 2014, www.scoop.co.nz/stories/AK1102/S00416/the-new-currency-of-man-lion-reds-man-points.htm.

109 **In another of Rozin's:** Paul Rozin et al., "Is Meat Male?," *Journal of Consumer Research* 39 (2012): 629–643.

109 **The notion that meat:** Adams, *Sexual Politics of Meat*, 48.

109 **A Tudor knight received:** Vaclav Smil, "Eating Meat: Evolution, Patterns, and Consequences," *Population and Development Review* 28 (2002): 599–639.

109 **Even much later:** Jeremy Rifkin, *Beyond Beef: The Rise and Fall of the Cattle Culture* (New York: Penguin Books, 1992), 249.

110 **If a woman dares steal:** John D. Speth, *Paleoanthropology and Archaeology of Big Game Hunting* (New York: Springer, 2010), 19.

110 **In Victorian times:** Joyce Stavick, "Love at First Beet: Vegetarian Critical Theory Meats Dracula," *Victorian Poetry* (1990): 24–25.

110 **Research shows that:** Jaime Mendiola et al., "Food Intake and Its Relationship with Semen Quality: A Case-Control Study," *Fertility and Sterility* 91 (2009): 812–818.

110 **What's more, if a guy's:** Shanna Swan et al., "Semen Quality of Fertile US Males in Relation to Their Mothers' Beef Consumption During Pregnancy," *Human Reproduction* 22 (2007): 1497–1502.

110 **A feminist, a writer:** Adams, *Sexual Politics of Meat*, 14.

110 **The British *Sunday Telegraph*:** Ibid.

111 **Adams explains:** Carol Adams, phone interview by author, June 26, 2014.

111 **One survey:** Peggy Reeves Sanday, *Female Power and Male Dominance: On the Origins of Sexual Inequality* (Cambridge, UK: Cambridge University Press, 1981), 66.

111 **"Back in the '80s:** Adams, interview.

111 **"Vegetables are for girls:** Hank Rothgerber, "Real Men Don't Eat (Vegetable) Quiche: Masculinity and the Justification of Meat Consumption," *Psychology of Men & Masculinity* 14 (2013): 363–375.

112 **Other researchers, too:** Rogers, "Beasts, Burgers, and Hummers."

112 **As Adams wrote:** Adams, *Sexual Politics of Meat*, 63.

113 **Already back then:** Katherine Clark, Salima Ikramb, and Richard Eversheda, "Organic Chemistry of Balms Used in the Preparation of Pharaonic Meat Mummies," *Proceedings of the National Academy of Sciences* 110 (2013).

113 **Psychological experiments show:** Jeremy Nicholson, "Does Playing Hard to Get Make You Fall in Love?," *Psychology Today*, April 19, 2012, accessed December 5, 2014, www.psychologytoday.com/blog/the-attraction-doctor/201204/does-playing-hard-get-make-you-fall-in-love.

113 **Meanwhile, the aristocracy:** Nerissa Russell, *Social Zooarchaeology: Humans and Animals in Prehistory* (Cambridge, UK: Cambridge University Press, 2012), 360 (see Travis R. Pickering and Manuel Domínguez-Rodrigo, "Chimpanzee Referents and the Emergence of Human Hunting," *The Open Anthropology Journal* 3 (2010): 111).

113 **When Henry IV:** Jack Cecil Drummond and Anne Wilbraham, *The Englishman's Food: A History of Five Centuries of English Diet* (London: J. Cape, 1958).

113 **One British "shopping" list:** Ibid.

114 **Back then if you:** Linda Civitello, *Cuisine and Culture: A History of Food and People* (Hoboken, NJ: John Wiley & Sons, 2011), 95.

114 **Owning a cow:** Rifkin, *Beyond Beef*, 28–31.

115 **Ergo, boiling is perfect:** Russell, *Social Zooarchaeology*, 360.

115 **Packs of swine:** Katharine Rogers, *Pork: A Global History* (London: Reaktion Books, 2012), 68.

115 **Pork was easier:** Roger Horowitz, *Putting Meat on the American Table: Taste, Technology, Transformation* (Baltimore, MD: Johns Hopkins University Press, 2006), 43–45.

116 **George Miller Beard:** George M. Beard, *Sexual Neurasthenia [Nervous Exhaustion]: Its Hygiene, Causes, Symptoms and Treatment with a Chapter on Diet for the Nervous* (New York: E. B. Treat, 1898), quoted in Adams, *Sexual Politics of Meat*, 54.

116 **In a book published:** Robert Byron Hinman and Robert Bernard Harris, *The Story of Meat* (Chicago: Swift & Company, 1942), 1.

116 **Even after the war:** Frank Gerrard, *Meat Technology: A Practical Textbook for Student and Butcher* (London: Leonard Hill, 1945), quoted in Adams, *Sexual Politics of Meat*, 48.

116 **Fiddes, a Scottish anthropologist:** Nick Fiddes, *Meat: A Natural Symbol* (London: Routledge, 1991).

Chapter 8: The Half-Crazed, Sour-Visaged Infidels, or Why Vegetarianism Failed in the Past

119 **Over two thousand years ago:** Colin Spencer, *The Heretic's Feast: A History of Vegetarianism* (Hanover, NH: University Press of New England, 1996), 47–48.

119 **There was even:** Christiane Joost-Gaugier, *Measuring Heaven* (Ithaca, NY: Cornell University Press, 2006), 49.

120 **According to one story:** Spencer, *The Heretic's Feast*, 47–48.

120 **To avoid such risks:** Ibid.

120 **He was quite:** Ibid., x.

120 **It also likely failed:** Ibid., 37.

121 **The wrestler Milo:** James C. Wharton, "Muscular Vegetarianism: The Debate over Diet and Athletic Performance in the Progressive Era," *Journal of Sport History* 8 (1981): 58.

121 **They thought that:** Irvin Rock, ed., *The Legacy of Solomon Asch: Essays in Cognition and Social Psychology* (Hillsdale, NJ: L. Erlbaum Associates, 1990), 101.

121 **Followers of Pythagoras:** Spencer, *The Heretic's Feast*, 86.

122 **But when Moses:** Deuteronomy 34:1–8.

122 **Yet back in its heyday:** "The Qumran Community," The Israel Museum, Jerusalem, accessed December 5, 2014, www.english.imjnet.org.il/page_1350.

122 **According to some:** Robert Eisenman, *James, the Brother of Jesus* (New York: Viking, 1997).

122 **After all, it was:** Genesis 2:9.

122 **Both sides seem:** Genesis 1:29.

122 **Or in the actual words:** Genesis 9:3.

122 **It's a permission granted:** Elizabeth Telfer, *Food for Thought: Philosophy and Food* (New York: Routledge, 2002), 168, 189.

123 **In the spring:** "The Digital Dead Sea Scrolls," The Israel Museum, Jerusalem, accessed December 5, 2014, http://dss.collections.imj.org.il/discovery.

123 **According to Eisenman:** Eisenman, *James, the Brother of Jesus*.

123 **James's disciples claimed:** Spencer, *The Heretic's Feast*, 112–113.

123 **In 68 CE when:** "The Qumran Community."

124 **Instead, meat eating:** Spencer, *The Heretic's Feast*, 163.

124 **In order to separate:** Ibid., 170.

124 **In the fourth century CE:** Ibid., 142.

125 **As Robert Eisenman once:** Robert Eisenman, e-mail message to author, February 22, 2015.

125 **The medieval heretics:** Spencer, *The Heretic's Feast*, 157.

125 **Even in mainstream:** Telfer, *Food for Thought*, 200.

125 **Just as for Pythagoras:** Genesis 1:26.

125 **That's what St. Thomas Aquinas:** Spencer, *The Heretic's Feast*, 176; Ibid., 121.

126 **He claimed only:** Richard W. Schwarz, *John Harvey Kellogg, M.D.: Pioneering Health Reformer* (Hagerstown, MD: Review and Herald, 2006), 143.

126 **He was controlling:** Margaret Puskar-Pasewicz, ed., *Cultural Encyclopedia of Vegetarianism* (Santa Barbara, CA: Greenwood, 2010), 147.

126 **He believed that:** Schwarz, *John Harvey Kellogg*, 147; Ibid., 60.

126 **What's more, he believed:** John Harvey Kellogg, *The Natural Diet of Man* (Arvada, CO: Coastalfields Press, 2006), 217.

126 **During one of his:** Ibid., 41.

126 **At the height:** Karen Iacobbo and Michael Iacobbo, *Vegetarian America: A History* (Westport, CT: Praeger, 2004), 130, see also Shefali Sharma, "Food and National Security: The Shuanghui-Smithfield Merger Revisited," Institute for Agriculture and Trade Policy, September 12, 2013, accessed December 2, 2014, www.iatp.org/blog/201309/food-and-national-security-the-shuanghui-smithfield-merger-revisited#sthash.oGoFpEkt.JqIBctu5.dpuf; Roberto Ferdman, "Americans Have Never Had So Few Options in

Deciding What Company Makes Their Meat," *Washington Post*, September 4, 2014, accessed December 3, 2014, www.washingtonpost.com/blogs/wonkblog/wp/2014/06/11/americans-have-never-had-so-few-options-in-deciding-what-company-makes-their-meat/.

126 **The diet offered:** Puskar-Pasewicz, *Cultural Encyclopedia*, 49.

126 **Thankfully, though:** Ibid.; Schwarz, *John Harvey Kellogg*, 50.

127 **Less than 20 percent:** "ABC News Poll: What's for Breakfast?" May 15, 2005, accessed December 5, 2014, http://abcnews.go.com/images/Politics/981a1Breakfast.pdf.

127 **Sylvester Graham wasn't:** Iacobbo and Iacobbo, *Vegetarian America*, 19–21.

127 **The youngest of seventeen:** Barbara Haber, *From Hardtack to Homefries: An Uncommon History of American Cooks and Meals* (New York: Free Press, 2002), 46; Iacobbo and Iacobbo, *Vegetarian America*, 14.

127 **Or, as Ralph Waldo Emerson:** Iacobbo and Iacobbo, *Vegetarian America*, 32.

128 **Even governments supported:** Ibid., 18.

128 **More vegetables:** Spencer, *The Heretic's Feast*, 296.

128 **The center of the vegetarian:** Puskar-Pasewicz, *Cultural Encyclopedia*, 233–236; Telfer, *Food for Thought*, 26.

128 **Some had been:** "Adjourned Conference of Vegetarians," The Truth Tester II (1847): 29, accessed December 5, 2014, www.ivu.org/congress/1847/conference2.html.

128 **That day over 150:** Ibid.

129 **You would get "pale:** "Personal Glimpses," *Literary Digest*, September 1, 1928, 43, quoted in Iacobbo and Iacobbo, *Vegetarian America*, 132.

129 **Vegetables were often:** Puskar-Pasewicz, *Cultural Encyclopedia*, 231.

129 **Even John Harvey Kellogg:** Iacobbo and Iacobbo, *Vegetarian America*, 129.

129 **Graham lectured that:** Linda Civitello, *Cuisine and Culture: A History of Food and People* (Hoboken, NJ: John Wiley & Sons, 2011), 237.

129 **Kellogg went even further:** Schwarz, *John Harvey Kellogg*, 28; Iacobbo and Iacobbo, *Vegetarian America*, 42.

129 **Meanwhile, Leo Tolstoy:** Spencer, *The Heretic's Feast*, 288.

130 **They were called:** Iacobbo and Iacobbo, *Vegetarian America*, 96, 156.

130 **Over one hundred years:** Ibid., 61.

130 **A vegetarian metropolis:** Ibid., 91.

130 **Even though these:** Spencer, *The Heretic's Feast*, 268.

131 **As one novelist:** Jojo Moyes, *The Girl You Left Behind* (New York: Pamela Dorman Books/Viking, 2012), 104.

131 **When a 1940 survey:** Iacobbo and Iacobbo, *Vegetarian America*, 158.

131 **In Britain, if you:** Nick Fiddes, *Meat: Natural Symbol* (London: Routledge, 1991), 28.

132 **Several of the dictator's:** Joachim C. Fest, *Hitler* (New York: Harcourt Brace Jovanovich, 1974), 535; Ian Kershaw, *Hitler 1889–1936: Hubris* (New York: W. W. Norton, 1999).

132 **He was quite:** Ibid.

132 **At the same time:** Spencer, *The Heretic's Feast*, 308.

132 **As *Rags* magazine:** Iacobbo and Iacobbo, *Vegetarian America*, 173.

133 **Europe was no India:** K. T. Achaya, *A Historical Dictionary of Indian Food* (New Delhi: Oxford University Press, 1998), 163.

Chapter 9: Why Giving Up Meat May Be Harder for Some of Us

135 **To the average American:** Matthew B. Ruby, "Vegetarianism: A Blossoming Field of Study," *Appetite* 58 (2012): 141–150; Pamela Goyan Kittler, Kathryn Sucher, and Marcia Nelms, *Food and Culture* (Belmont, CA: Wadsworth, 2012), 3.

136 **An American comedian:** Robert Byrne, *The 2,548 Best Things Anybody Ever Said* (New York: Simon & Schuster, 2006), 97.

136 **Recent estimates show:** Margaret Puskar-Pasewicz, ed., *Cultural Encyclopedia of Vegetarianism* (Santa Barbara, CA: Greenwood, 2010), 234; Hank Rothgerber, "Real Men Don't Eat (Vegetable) Quiche: Masculinity and the Justification of Meat Consumption," *Psychology of Men & Masculinity* 14 (2013) (see also Jeremy Nicholson, "Does Playing Hard to Get Make You Fall in Love?," *Psychology Today*, April 19, 2012, accessed December 5, 2014, www.psychologytoday.com/blog/the-attraction-doctor/201204/does-playing-hard-get-make-you-fall-in-love); Ruby, "Vegetarianism."

136 **Che Green, executive director:** Mat Thomas, "The Road to Vegetopia: (Re) Imaging the Future of Food," *VegNews* March/April (2009): 36, quoted in Greg Goodale and Jason Edward Black, eds., *Arguments About Animal Ethics* (Lanham, MD: Lexington Books, 2010).

136 **It seems that wherever:** Puskar-Pasewicz, *Cultural Encyclopedia*, 243–24.

137 **The most common:** Ruby, "Vegetarianism."

137 **As *Saturday Night Live*:** Richard Zera, *Business Wit & Wisdom* (Washington, DC: Beard Books, 2005), 54.

137 **They may watch:** Ruby, "Vegetarianism."

137 **Psychologists say that:** Nick Fox and Katie Ward, "You Are What You Eat? Vegetarians, Health and Identity," *Social Science & Medicine* 66 (2008): 2585–2595.

138 **When I ask her about:** Kate Jacoby, interview by author, Philadelphia, PA, October 15, 2013.

138 **The author of the:** Scott Gold, *The Shameless Carnivore: A Manifesto for Meat Lovers* (New York: Broadway Books, 2008), 150.

138 **Back in the 1940s:** Hyman S. Barahal, "The Cruel Vegetarian," *Psychiatric Quarterly* Supplement 20 (1946), quoted in Ruby, "Vegetarianism."

138 **Twenty-first-century studies:** Ruby, "Vegetarianism."

138 **In one experiment:** Jean Anthelme Brillat-Savarin, *The Physiology of Taste or, Meditations on Transcendental Gastronomy* (New York: Everyman's Library, 2009), 132.

139 **The results were:** Kittler, Sucher, and Nelms, *Food and Culture*.

139 **Psychologists have found:** Matthew Ruby and Steven Heine, "Meat, Morals, and Masculinity," *Appetite* 56 (2011): 447–450.

139 **This perceived lack:** "Why Some Meat-Eaters Won't Date Vegetarians," *Cosmopolitan*, July 9, 2012, accessed December 7, 2014, www.cosmopolitan.com/health-fitness/news/a10375/dating-vegetarians/.

139 **The source: a study:** Annie Potts and Jovian Parry, "Vegan Sexuality: Challenging Heteronormative Masculinity Through Meat-free Sex," *Feminism & Psychology* (2010): 53–68.

139 **On Internet forums:** Ibid.

139 **According to one experiment:** Jan Havlicek and Pavlina Lenochova, "The Effect of Meat Consumption on Body Odour Attractiveness," *Chemical Senses* 31 (2006): 747–752.

139 **If you send:** Klaus Petzke, Heiner Boeing, and Cornelia Metges, "Choice of Dietary Protein of Vegetarians and Omnivores Is Reflected in Their Hair Protein 13C and 15N Abundance," *Rapid Communications in Mass Spectrometry* 19 (2005): 1392–1400.

140 **By placing electrodes:** Jessica Stockburger et al., "Vegetarianism and Food Perception: Selective Visual Attention to Meat Pictures," *Appetite* 52 (2009): 513–516.

140 **A study conducted:** Fiona Breen, Robert Plomin, and Jane Wardle, "Heritability of Food Preferences in Young Children," *Physiology & Behavior* 88 (2006): 443–447.

140 **Another study:** PS Prado-Lima et al., "Human Food Preferences Are Associated with a 5-HT2A Serotonergic Receptor Polymorphism," *Molecular Psychiatry* 11 (2006): 889–891.

140 **Twin studies show:** Lucy Cooke, Claire Haworth, and Jane Wardle, "Genetic and Environmental Influences on Children's Food Neophobia," *American Journal of Clinical Nutrition* 86 (2007): 428–433.

140 **Neophobes are the people:** Hely Tuorila et al., "Food Neophobia Among the Finns and Related Responses to Familiar and Unfamiliar Foods," *Food Quality and Preference* 12 (2001): 29–37.

141 **They are also less:** Trevor Lunn et al., "Does Personality Affect Dietary Intake?," *Nutrition* 30 (2014): 403–409.

141 **Research shows that:** Catharine Gale et al., "IQ in Childhood and Vegetarianism in Adulthood: 1970 British Cohort Study," *BMJ* 334 (2007): 245.

141 **"We want meat eaters:** Evelyn Kimber, interview by author, Boston, MA, October 26, 2013.

142 **A recent experiment:** David Gal and Derek D. Rucker, "When in Doubt, Shout!: Paradoxical Influences of Doubt on Proselytizing," *Psychological Science* (2010): 1701–1707.

142 **(those who went:** Terry McConnell, "Irresistible Force Versus Immovable Object," *Edmonton Journal*, March 20, 2005, D2.

142 **Since for vegans:** Hank Rothgerber, "Horizontal Hostility Among Non-Meat Eaters," *PLoS ONE* 9 (2014).

142 **That is also why:** Ibid.

142 **In experiments, putting:** Ibid.

143 **Kristin Lajeunesse, an author:** Kristin Lajeunesse, interview by author, Boston, MA, October 26, 2013.

143 **But those omnivores:** Hank Rothgerber, "Efforts to Overcome Vegetarian-Induced Dissonance Among Meat Eaters," *Appetite* 79 (2014): 32–41.

143 **A common strategy:** Ibid.

143 **As a journalist:** Verena Besso, "Reason, and the Rights of Animals," *Toronto Star*, October 4, 1990, A25.

144 **Bastian was raised:** Brock Bastian, phone interview by author, August 27, 2013.

144 **In one of his experiments:** Brock Bastian et al., "Don't Mind Meat? The Denial of Mind to Animals Used for Human Consumption," *Personality and Social Psychology Bulletin* 38 (2012): 247.

144 **"Thinking of a cow:** Bastian, interview.

144 **This time he made:** Boyka Bratanova, Steve Loughnan, and Brock Bastian, "The Effect of Categorization as Food on the Perceived Moral Standing of Animals," *Appetite* 57 (2011): 193–196.

145 **While men are more:** Rothgerber, "Real Men Don't Eat."

145 **Would we be so:** Bernard Shaw, *Shaw: An Autobiography 1856–1898; Selected from His Writings*, ed. Stanley Weintraub (New York: Weybright and Talley, 1969), 92.

145 **Experiments show:** Melanie Joy, *Why We Love Dogs, Eat Pigs, and Wear Cows: An Introduction to Carnism* (San Francisco: Conari Press, 2010), 121.

145 **Fifty-eight billion:** "Meat Indigenous, Chicken," FAOstat, accessed December 7, 2014, http://faostat.fao.org/site/569/DesktopDefault.aspx?PageID=569#ancor.

145 **Returning to our:** Rothgerber, "Efforts to Overcome."

146 **In one Canadian survey:** Ruby, "Vegetarianism."

146 **In another poll:** Rothgerber, "Efforts to Overcome."

146 **This means that:** Alan Beardsworth and Teresa Keil, *Sociology on the Menu: An Invitation to the Study of Food and Society* (London: Routledge, 1997), 225.

146 **This particular technique:** Rothgerber, "Efforts to Overcome."

146 **When people go "veg":** Daniel Fessler et al., "Disgust Sensitivity and Meat Consumption: A Test of an Emotivist Account of Moral Vegetarianism," *Appetite* 41 (2003): 31–41.

147 **When you cook it:** Richard Landau, interview by author, Philadelphia, PA, October 15, 2013.

147 **Approximately 45 percent:** David T. Neal, Wendy Wood, and Jeffrey M. Quinn, "Habits—A Repeat Performance," *Current Directions in Psychological Science* 15 (2006): 198–202.

147 **Same goes for food:** Brian Wansink, *Mindless Eating: Why We Eat More Than We Think* (New York: Bantam Books, 2006), 94.

147 **We like our habits:** Géraldine Coppin and David Sander, "The Flexibility of Chemosensory Preferences," *Neuroscience of Preference and Choice* (2011): 267.

147 **"Many people wonder:** Kimber, interview.

148 **Lack of cooking skills:** Jennie Yabroff, "No More Sacred Cows," *Newsweek*, January 11, 2010, 66.

148 **In surveys they:** Kenneth Menzies and Judy Sheeshka, "The Process of Exiting Vegetarianism: An Exploratory Study," *Canadian Journal of Dietetic Practice and Research* 73 (2012): 165.

148 **Things like turnips:** Landau, interview.

148 **Says Landau:** Ibid.

148 **A typical ex-vegetarian:** Menzies and Sheeshka, "The Process of Exiting Vegetarianism," 167.

Chapter 10: Dog Skewers, Beef Burgers, and Other Weird Meats

151 **This is a traditional:** Calvin W. Schwabe, *Unmentionable Cuisine* (Charlottesville: University Press of Virginia, 1979), 174.

151 **It was in the Ituri Forest:** Richard Wrangham, phone interview by author, December 1, 2013.

151 **Wrangham, primatologist Elizabeth Ross:** Wrangham, interview.

153 **To some Somali tribes:** Frederick J. Simoons, *Eat Not This Flesh: Food Avoidances from Prehistory to the Present* (Madison: University of Wisconsin Press, 1994), 161, 261–262.

153 **Meanwhile in Asia:** Anthony Podberscek, "Good to Pet and Eat: The Keeping and Consuming of Dogs and Cats in South Korea," *Journal of Social Issues* 65 (2009): 617.

153 **Dogs, according:** Ibid., 623.

154 **Professor Alan Reilly:** Jethro Mullen, "Horsemeat Found in Hamburgers in Britain and Ireland," CNN Wire, January 16, 2013.

154 **They left archaeologists:** Finbar McCormick, "Ritual Feasting in Iron Age Ireland," in *Relics of Old Decency: Archaeological Studies in Later Prehistory*, ed. Gabriel Cooney et al. (Dublin, Ireland: Wordwell, 2009), 410.

154 **In prehistoric times:** Chris Otter, "Hippophagy in the UK: A Failed Dietary Revolution," *Endeavour* 35 (2011): 81.

154 **According to the Bible:** Leviticus 11:8.

154 **To make things worse:** Schwabe, *Unmentionable Cuisine*, 158; Simoons, *Eat Not This Flesh*, 183–184.

154 **In the end, rather:** Simoons, *Eat Not This Flesh*, 187–188.

154 **Even faced with:** Ibid., 188.

154 **A ninth-century Irish:** Ibid., 187.

154 **longer than was:** Byrne Fone, *Homophobia: A History* (New York: Metropolitan Books, 2000), 124.

155 **Their carcasses would:** Otter, "Hippophagy in the UK," 84.

155 **In the medical press:** Ibid., 83.

155 **According to historians:** Ibid.

155 **The venue was magnificent:** *Daily Phoenix* (Columbia, SC), July 26, 1865, accessed January 5, 2015, http://chroniclingamerica.loc.gov.

156 **The guests applauded:** Tom Hughes, "Falling at the First," *Marylebone Journal*, accessed January 15, 2015, http://marylebonejournal.com/history/falling-at-the-first.

156 **He dined on Japanese:** Jerry Hopkins, *Extreme Cuisine: The Weird & Wonderful Foods That People Eat* (Singapore: Periplus, 2004), 11–12.

156 **A few days after:** Hughes, "Falling at the First."

156 **In 1879, the:** Otter, "Hippophagy in the UK."

156 **As one historian said:** Ibid.

156 **Although horsemeat was:** Sylvain Leteux, "Is Hippophagy a Taboo in Constant Evolution?" *Menu: Journal of Food and Hospitality* (2012): 1–13.

157 **China is the most:** Ibid.

157 **According to the Kazakhs:** Marsha Levine, "Eating Horses: The Evolutionary Significance of Hippophagy," *Antiquity* 72 (1998): 90–100.

157 **Yet anthropologist:** Daniel Fessler, e-mail message to author, August 19, 2014.

157 **According to a theory:** Clyde Manwell and Ann Baker, "Domestication of the Dog: Hunter, Food, Bed-warmer, or Emotional Object?," *Zeitschrift für Tierzüchtung und Züchtungsbiologie* 101 (1984): 241–256.

157 **By the Bronze Age:** Sarah Knight and Harold Herzog, *New Perspectives on Human-Animal Interactions: Theory, Policy and Research* (Malden, MA: Wiley-Blackwell, 2009), 616.

157 **The Ancient Greeks:** Sophia Menache, "Dogs: God's Worst Enemies?," *Society and Animals* 5 (1997): 27.

157 **Even the early:** Earley Vernon Wilcox and Clarence Beaman Smith, *Farmer's Cyclopedia of Live Stock* (New York: Orange Judd Company, 1908), 691.

157 **Today, about sixteen million:** Podberscek, "Good to Pet and Eat."

158 **As long as it is properly:** Jonathan Safran Foer, "Let Them Eat Dog," *Wall Street Journal*, October 31, 2009.

158 **South Koreans, the biggest fans:** Podberscek, "Good to Pet and Eat."

158 **Also, because dog:** Ibid.

158 **One thing the campaign:** Ibid.

158 **The lack of consistency:** Ibid.

158 **If South Koreans:** Ibid.

159 **And good for:** "Kawior i Ślimaki. Polska to "Luksusowy" Eksporter," TVN24, July 12, 2011, accessed January 15, 2015, http://tvn24bis.pl/wiadomosci-gospodarcze,71/kawior-i-slimaki-polska-to-luksusowy-eksporter,177782.html.

159 **The !Kung bushmen:** Frederic Simoons, "Traditional Use and Avoidance of Foods of Animal Origin," *BioScience* 28 (1978): 178–184.

159 **North Americans are:** Daniel Fessler and Carlos David Navarrete, "Meat Is Good to Taboo," *Journal of Cognition and Culture* 3 (2003): 1–40.

160 **The majority of:** Podberscek, "Good to Pet and Eat."

160 **The worth of:** Ibid.

160 **Surprisingly, pet owners:** Ibid.

160 **In Melanesia, pigs are:** Simoons, *Eat Not This Flesh*.

160 **They can be taught:** "Pigs Smarter Than They Look," *Lewiston Daily Sun* (Maine), February 21, 1984, 23.

161 **At Pennsylvania State:** Miguel Helft, "Pig Video Arcades Critique Life in the Pen," *Wired*, June 6, 1997, accessed January 15, 2015, http://archive.wired.com/science/discoveries/news/1997/06/4302.

161 **It appears that:** Liz Shankland, "Are They the Next Pig Thing?," *Western Mail* (Cardiff, Wales), November 30, 2004, 11.

161 **Cows may not:** "The Hidden Lives of Cows," PETA, accessed January 15, 2015, www.peta.org/issues/animals-used-for-food/hidden-lives-of-cows.aspx.

161 **For example, they:** Ingvar Ekesbo, *Farm Animal Behaviour: Characteristics for Assessment of Health and Welfare* (Wallingford, UK: CABI, 2011), 116.

162 **It was only in 1859:** Marvin Harris, *The Sacred Cow and the Abominable Pig: Riddles of Food and Culture* (New York: Simon & Schuster, 1987), 69.

162 **Undercooked beef is:** Ibid.

162 **Anthropologist Marvin Harris:** Ibid., 71.

162 **If the Jews and Muslims:** Ibid., 70–73.

163 **The earliest Vedas:** Marvin Harris, "India's Sacred Cow," *Human Nature* (1978): 28–36.

163 **"The farmers who decided:** Ibid.

163 **It's been calculated:** Ibid.

163 **And so, as Harris:** Ibid.

164 **Some such tribes:** Simoons, *Eat Not This Flesh*, 262.

164 **These taboos acted:** Richard Wrangham, e-mail message to author, August 22, 2013.

164 **Other scientists agree:** Nigel Barber, "Why Are Some Animals Considered Unclean?" *Psychology Today*, February 10, 2011, accessed January 15, 2015, www.psychologytoday.com/blog/the-human-beast/201102/why-are-some-animals-considered-unclean.

Chapter 11: The Pink Revolution, or How Asia Is Getting Hooked on Meat, Fast

167 **In fact, their urine:** Elizabeth Soumya, "Sacred Cows and Politics of Beef in India," Aljazeera, April 20, 2014, accessed January 6, 2015, www.aljazeera.com/indepth/features/2014/04/india-bjp-piggybacks-cow-milk-votes-2014417142154567121.html.

168 **What's more, India:** "India's Agricultural Exports Climb to Record High," USDA, August 29, 2014, accessed January 6, 2015, www.fas.usda.gov/data/india-s-agricultural-exports-climb-record-high.

168 **Granted, India still:** Mark Eisler et al., "Agriculture: Steps to Sustainable Livestock," *Nature* 507 (2014): 32–34.

168 **By 2030:** Timothy Robinson and Francesca Pozzi, "Mapping Supply and Demand for Animal-Source Foods to 2030," FAO, 2011, accessed January 6, 2015, www.fao.org/docrep/014/i2425e/i2425e00.pdf.

168 **There are five recognized:** Benjamin Caballero and Barry Popkin, eds., *The Nutrition Transition: Diet and Disease in the Developing World* (Amsterdam, Netherlands: Academic Press, 2002).

168 **One study showed:** Richard York and Marcia Hill Gossard, "Cross-national Meat and Fish Consumption: Exploring the Effects of Modernization and Ecological Context," *Ecological Economics* 48 (2004): 293–302.

169 **As late as 1939:** Vaclav Smil and Kazuhiko Kobayashi, *Japan's Dietary Transition and Its Impacts* (Cambridge, MA: MIT Press, 2012), 49.

169 **Today, the daily meat:** "Livestock and Fish Primary Equivalent," FAOstat, 2011, accessed January 7, 2015, http://faostat.fao.org/site/610/Desktop Default.aspx?PageID=610#ancor.

169 **Growing livestock takes:** Katarzyna J. Cwiertka, *Modern Japanese Cuisine: Food, Power and National Identity* (Chicago: University of Chicago Press,

2007), 24–27; see also Hank Rothgerber, "Real Men Don't Eat (Vegetable) Quiche: Masculinity and the Justification of Meat Consumption," *Psychology of Men & Masculinity* 14 (2013).

169 **That year, on January 24:** R. Kenji Tierney and Emiko Ohnuki-Tierney, "Anthropology of Food," in *The Oxford Handbook of Food History*, ed. Jeffrey M. Pilcher (Oxford, UK: Oxford University Press, 2012), 128.

170 **Over just five years:** Cwiertka, *Modern Japanese Cuisine*, 33; Ibid., 152.

170 **Back then, typical:** Ibid., 63–64.

170 **The words of Den Fujita:** Jeremy Rifkin, *Beyond Beef: The Rise and Fall of the Cattle Culture* (New York: Penguin Books, 1992), 271.

171 **as Anjanappa explains:** Ajath Anjanappa, interview by author, Bengaluru, India, January 6, 2014.

171 **Unbeknownst to many:** PK Krishnakumar, "India's Beef Exports Rise 31% in 2013–14," *Economic Times* (India), June 25, 2014.

172 **A big chunk:** "Exports from India of Buffalo Meat," APEDA Agri Exchange, 2014–2015, accessed January 7, 2015, http://agriexchange.apeda.gov.in/product_profile/exp_f_india.aspx?categorycode=0401.

172 **Instead, water buffaloes:** Pratiksha Ramkumar, "Beef Exports up 44% in 4 Years, India Top Seller," *Times of India*, April 1, 2013.

172 **According to a local:** Cithara Paul, "UPA's Pink Revolution Makes India World's Biggest Beef Exporter," *Sunday Standard* (India), February 9, 2014; Soumya, "Sacred Cows and Politics."

172 **As Anjanappa tells me:** Anjanappa, interview.

172 **Meanwhile, even though:** Benjamin Caballero, Lindsay Allen, and Andrew Prentice, eds., *Encyclopedia of Human Nutrition* (Amsterdam, Netherlands: Elsevier/Academic Press, 2005), 98.

172 **Today, devout Hindus:** Rifkin, *Beyond Beef*, 37; Colin Spencer, *The Heretic's Feast: A History of Vegetarianism* (Hanover, NH: University Press of New England, 1996, 77; see also Colin Spencer, *The Heretic's Feast: A History of Vegetarianism* (Hanover, NH: University Press of New England, 1996), 47–48.

173 **Until recently, killing:** Rifkin, *Beyond Beef*, 37.

173 **The fact that:** Soumya, "Sacred Cows and Politics."

173 **Vegetarianism in India:** Ludwig Alsdorf, *The History of Vegetarianism and Cow-Veneration in India* (London: Routledge, 2010).

173 **Although he was born:** Mahatma Gandhi, *An Autobiography: The Story of My Experiments with Truth* (Waiheke Island, New Zealand: Floating Press, 2009), 47–49.

174 **He admitted in his:** Ibid., 90.

174 **But in time, Gandhi:** Ibid.

174 **The protein myth:** Teja Lele Desai, "How They Turned Vegetarians and for Good," *Times of India*, April 26, 2014.

175 **Just think about it:** Margaret Puskar-Pasewicz, *Cultural Encyclopedia of Vegetarianism* (Santa Barbara, CA: Greenwood, 2010), 136.

175 **A prominent Indian food:** K. T. Achaya, *A Historical Dictionary of Indian Food* (New Delhi: Oxford University Press, 1998), 263.

175 **"Beef is a symbol:** "HCU Students Eating Beef, Organized ASA," YouTube, September 24, 2012, accessed January 7, 2015, www.youtube.com/watch?v=AB-Z99unUMU.

175 **It opposes the caste:** Rick Dolphijn, "Capitalism on a Plate: The Politics of Meat Eating in Bangalore, India," *Gastronomica: The Journal of Critical Food Studies* 6 (2006): 52–59.

175 **Before the elections:** Shilpa Kannan, "India's Elections Spark Debate on Beef Exports," BBC, May 4, 2014, accessed January 8, 2015, www.bbc.com/news/business-27251802.

176 **In a country where:** Vaclav Smil, "Eating Meat: Evolution, Patterns, and Consequences," *Population and Development Review* 28 (2002): 599–639.

176 **Since the 1980s:** Mindi Schneider and Shefali Sharma, "China's Pork Miracle? Agribusiness and Development in China's Pork Industry," Institute for Agriculture and Trade Policy, February 2014, http://www.iatp.org/files/2014_03_26_PorkReport_f_web.pdf.

176 **China is already:** Shefali Sharma, "The Need for Feed: China's Demand for Industrialized Meat and Its Impacts," Institute for Agriculture and Trade Policy, February 2014, http://www.iatp.org/files/2014_03_26_FeedReport_f_web.pdf.

176 **As anthropologist Yunxiang Yan:** Yunxiang Yan, "McDonald's in Beijing: The Localization of Americana," in *Golden Arches East: McDonald's in East Asia*, ed. James L. Watson (Stanford, CA: Stanford University Press, 1997), 42.

177 **The Chinese ate:** Frederick Simoons, *Food in China: A Cultural and Historical Inquiry* (Boca Raton, FL: CRC Press, 1991), 295.

177 **By the sixth century:** John Kieschnick, "Buddhist Vegetarianism in China," in *Of Tripod and Palate: Food, Politics and Religion in Traditional China*, ed. Roel Sterckx (New York: Palgrave Macmillan, 2005), 191–203.

178 **The rich of China:** Roel Sterckx, "Food and Philosophy in Early China," in *Of Tripod and Palate*, ed. Sterckx, 39.

178 **To rise through:** Kieschnick, "Buddhist Vegetarianism in China," 204.

178 **Meat avoiders were:** Vincent Goossaert, "The Beef Taboo and the Sacrificial Structure of Late Imperial Chinese Society," in *Of Tripod and Palate*, ed. Sterckx, 239.

178 **To make the character:** Katharine Rogers, *Pork: A Global History* (London: Reaktion Books, 2012), 91.

179 **In time I came:** Ibid.

179 **Every other pig:** Sharma, "The Need for Feed."

179 **The Communist government:** Fred Gale, Daniel Marti, and Dinghuan Hu, "China's Volatile Pork Industry," USDA, February 2012, accessed January 8, 2015, www.ers.usda.gov/media/262067/ldpm21101_1_.pdf.

179 **There are already:** Meena Daivadanam, "Lifestyle Change in Kerala, India: Needs Assessment and Planning for a Community-Based Diabetes Prevention Trial," *BMC Public Health* 13 (2013).

179 **In the twin city:** Constanze Weigl, "Lifestyle Diseases in India—The Management of Type 2 Diabetes Mellitus (T2DM) in Kerala," *Viennese Ethnomedicine Newsletter* 13 (2011): 40–47.

180 **The waists of Asia:** Shan Juan, "Chinese Consume Too Much Food from Animals, Experts Say," *China Daily*, February 13, 2014, 4.

180 **One hundred million:** Schneider and Sharma, "China's Pork Miracle?"

180 **In March 2013:** Nicola Davison, "China Loves Pork Too Much," *Guardian*, March 23, 2013, accessed January 8, 2015, www.theguardian.com/commentisfree/2013/mar/23/china-loves-pork-pig-carcasses.

180 **There was the:** Chendong Pi, Zhang Rou, and Sarah Horowitz, "Fair or Fowl? Industrialization of Poultry Production in China," Institute for Agriculture and Trade Policy, February 2014.

180 **Take ractopamine:** Schneider and Sharma, "China's Pork Miracle?"

180 **But when the Chinese:** Ibid.

180 **China has a mere:** "Arable Land (Hectares per Person)," World Bank, 2014, accessed January 8, 2015, http://data.worldbank.org/indicator/AG.LND.ARBL.HA.PC.

181 **In the last decade:** Nadia Arumugam, "Shuanghui-Smithfield Deal Means China Wants American Pork: Surely This Is a Good Thing?," *Forbes*, June 6, 2013, accessed January 8, 2015, www.forbes.com/sites/nadiaarumugam/2013/06/06/shuanghui-smithfield-deal-means-china-wants-american-pork-surely-this-is-a-good-thing/.

181 **While the Chinese:** Tom Philpott, "Are We Becoming China's Factory Farm?," *Mother Jones*, November 1, 2013, 72.

181 **A lot of it comes:** "Changes of Soybean Imports from Different Countries in 2012," CnAgri, January 30, 2013, accessed January 9, 2015, http://en.cnagri.com/news/insight/20130130/296455.html.

181 **Already over 80 percent:** Sharma, "The Need for Feed," 24.

181 **A slice of Brazil:** Ibid., 26.

181 **Ninety-nine percent of the:** Beth Hoffman, "How Increased Meat Consumption in China Changes Landscapes Across the Globe," *Forbes*, March 26, 2014, accessed January 9, 2015, www.forbes.com/sites/bethhoffman/2014/03/26/how-increased-meat-consumption-in-china-changes-landscapes-across-the-globe/.

181 **In Argentina's soy-growing:** Sharma, "The Need for Feed," 29.

CHAPTER 12: THE FUTURE OF OUR MEAT-BASED DIETS

183 **Already, 33 percent:** "Livestock a Major Threat to Environment," UN Food and Agriculture Organization (FAO), November 26, 2006, accessed January 9, 2015, www.fao.org/newsroom/en/News/2006/1000448/index.html.

183 **If the 9.3 billion people:** World meat production in 2014 was about 256 million tons, see Rob Cook, "World Meat Production (1960–2014)," Beef2Live, accessed January 9, 2015, http://beef2live.com/story-world-meat-production-1960-2014-89-111818; American meat consumption was 125 kg (275.5 pounds) per person per year in 2007, see Mark C. Eisler et al., "Agriculture: Steps to Sustainable Livestock," *Nature* 507 (2014); world milk production is projected to increase from 580 million tons in 2001 to 1,043 million tons in 2050, see "Livestock a Major Threat to Environment," FAO; American dairy consumption is about 630 pounds per person per year, see Allison Aubrey, "The Average American Ate (Literally) a Ton This Year," December 31, 2011, NPR, accessed January 9, 2015, www.npr.org/blogs/thesalt/2011/12/31/144478009/the-average-american-ate-literally-a-ton-this-year.

183 **Even in many rich:** James Galloway et al., "International Trade in Meat: The Tip of the Pork Chop," *Ambio* 36 (2007): 622–629.

184 **In 2013 it was:** "Food Outlook," FAO, June 2013, accessed January 9, 2015, www.fao.org/docrep/018/al999e/al999e.pdf.

184 **If the current growth:** Shefali Sharma, "The Need for Feed: China's Demand for Industrialized Meat and Its Impacts," Institute for Agriculture and Trade Policy, February 2014, 9, http://www.iatp.org/files/2014_03_26_FeedReport_f_web.pdf.

184 **In general, to feed:** "How to Feed the World in 2050," FAO, accessed January 9, 2015, www.fao.org/wsfs/forum2050/wsfs-forum/en/.

184 **To grow a pound:** David Pimentel and Marcia Pimentel, "Sustainability of Meat-Based and Plant-Based Diets and the Environment," *American Journal of Clinical Nutrition* 78 (2003): 660S–663S.

184 **In the US, livestock:** Vaclav Smil, "Eating Meat: Evolution, Patterns, and Consequences," *Population and Development Review* 28 (2002) (see Amanda

Radke, "New Beef, Pork Names Don't Do Us Any Special Favors," *Beef*, June 27, 2013).

184 **A pound of beef:** "Meat Atlas," Friends of the Earth Europe, January 2014, accessed January 10, 2015, www.foeeurope.org/sites/default/files/publications/foee_hbf_meatatlas_jan2014.pdf.

184 **In about a decade:** Ibid.

184 **Of all greenhouse:** "Major Cuts of Greenhouse Gas Emissions from Livestock Within Reach," FAO, September 26, 2013, accessed January 10, 2015, www.fao.org/news/story/en/item/197608/icode/.

184 **If this number doesn't:** "Reducing Transport Greenhouse Gas Emissions: Trends & Data 2010," OECD/ITF, accessed January 10, 2015, www.internationaltransportforum.org/Pub/pdf/10GHGTrends.pdf.

184 **If we do nothing:** Elke Stehfest et al., "Climate Benefits of Changing Diet," *Climatic Change* 95 (2009): 83–102.

184 **To have a good chance:** Frederik Hedenus, Stefan Wirsenius, and Daniel J. A. Johansson, "The Importance of Reduced Meat and Dairy Consumption for Meeting Stringent Climate Change Targets," *Climate Change* 124 (March 28, 2014): 79–91.

185 **"It's very close:** Hanni Rützler, interview by author, London, August 5, 2013.

185 **"Many TV crews**: Anon van Essen, interview by author, Maastricht, May 12, 2014.

186 **And yet the mastermind:** Mark Post, interview by author, London, August 5, 2013.

186 **In theory, lab-grown:** Marta Zaraska, "Lab-Grown Beef Taste Test: 'Almost' Like a Burger," *Washington Post*, August 5, 2013, 1.

187 **Even though cultured:** Brandon Griggs, "How Test-Tube Meat Could Be the Future of Food," CNN Wire, April 30, 2014.

187 **"These cells are dead:** Van Essen, interview.

187 **"I think people are:** Clément Scellier, interview by author, Paris, February 27, 2014.

187 **"That's just bad quality:** Bastien Rabastens, interview by author, Paris, February 27, 2014.

188 **Yet two billion other:** Claire MacEvilly, "Bugs in the System," *Nutrition Bulletin* 25 (2000): 267–268.

188 **In Uganda, a pound:** Arnold van Huis, "Potential of Insects as Food and Feed in Assuring Food Security," *Annual Review of Entomology* 58 (2013): 563–583.

188 **Plenty of species:** Ibid.

188 **Because they are:** Ibid.

188 **Arnold van Huis, professor:** Arnold van Huis, interview by author, Wageningen, May 13, 2014.

189 **Already, one out of:** Wim Verbeke, "Profiling Consumers Who Are Ready to Adopt Insects as a Meat Substitute in a Western Society," *Food Quality and Preference* 39 (2015): 147–155.

189 **In the States:** "Defect Levels Handbook," FDA, accessed January 10, 2015, www.fda.gov/food/guidanceregulation/guidancedocumentsregulatory information/sanitationtransportation/ucm056174.htm.

189 **We could also:** "Insects au Gratin: LSBU Academics 3D Print Edible Insects," London South Bank University, February 11, 2014, accessed January 10, 2015.

190 **Korteweg, the "butcher":** Jaap Korteweg, interview by author, The Hague, May, 14, 2014.

190 **"Come and taste it:** Ibid.

190 **When Ferran Adrià:** Ibid.

191 **When in 2013:** Stephanie Strom, "Fake Meats, Finally, Taste Like Chicken," *New York Times*, April 3, 2014, 1.

191 **What's more, nutritionally:** Ibid.

191 **"This smells like Band-Aids":** "Pilot," *Breaking Bad*, AMC, New York, January 20, 2008.

191 **The global meat:** "Meat Substitutes Market Worth $4,622.4 Million by 2019," PR Newswire, June 3, 2014, http://www.prnewswire.com/news-releases/meat-substitutes-market-worth-46224-million-by-2019-261662621.html.

191 **Compared to real meat:** Glenn Zorpette, "The Better Meat Substitute," IEEE Spectrum, June 3, 2013, accessed January 10, 2015, http://spectrum.ieee.org/energy/environment/the-better-meat-substitute.

191 **Saying it tastes:** Brian Wansink, Mitsuru Shimizu, and Adam Brumberg, "Dispelling Myths About a New Healthful Food Can Be More Motivating Than Promoting Nutritional Benefits: The Case of Tofu," *Eating Behaviors*, 15 (2014): 318–320.

191 **"We make vegan ham:** Ingrid Newkirk, phone interview by author, June 17, 2014.

191 **In the States:** Zorpette, "The Better Meat Substitute."

192 **The first would be:** "Global Food Losses and Food Waste—Extent, Causes and Prevention," FAO, 2011, accessed January 10, 2015, www.fao.org/docrep/014/mb060e/mb060e.pdf.

192 **Half of unused meat:** Ibid.

192 **In Europe, it has:** Ffion Lloyd-Williams et al., "Estimating the Cardiovascular Mortality Burden Attributable to the European Common Agricultural Policy on Dietary Saturated Fats," *Bulletin of the World Health Organization* 86 (2008): 497–576.

192 **A "fat tax":** Torben Jørgensen et al., "The Danish Fat Tax—a Story of Political Incompetence?," *European Journal of Public Health* 24 (2014).

193 **In a soft voice:** Peter Singer, phone interview by author, June 23, 2014.

193 **Singer himself says:** Ibid.

193 **As Paul Rozin once:** Paul Rozin, interview by author, Paris, July 20, 2014.

194 **At the time of writing:** "Meat Atlas," Friends of the Earth Europe.

194 **In the States:** Eliza Barclay, "Why There's Less Red Meat on Many American Plates," NPR, June 27, 2012, accessed January 11, 2015, www.npr.org/blogs/thesalt/2012/06/27/155837575/why-theres-less-red-meat-served-on-many-american-plates.

194 **In Germany that number:** Georgi Gyton, "Spotlight on Flexitarianism," September 3, 2014, accessed January 11, 2015, www.globalmeatnews.com/Analysis/Spotlight-on-flexitarianism-are-consumers-cutting-their-meat-intake.

194 **Moreover, polls show:** Nick Cooney, *Veganomics: The Surprising Science on What Motivates Vegetarians, from the Breakfast Table to the Bedroom* (New York: Lantern Books, 2014).

194 **Lord Stern, former vice president:** Steve Dubé, "Welsh Red Meat Is Better for the Environment Than Other Products: UK Livestock Farming Accounts for only 2.9% of Climate Change Gases," *Western Mail* (Cardiff, Wales), November 17, 2009, 3.

194 **But Morgaine Gaye:** Morgaine Gaye, interview by author, London, July 14, 2014.

194 **"I don't think we will:** Ibid.

194 **She also tells me:** Ibid.

195 **Or "cultured," as Andras:** Andras Forgacs, phone interview by author, September 25, 2014.

195 **As Forgacs once told me:** Ibid.

195 **That's why Modern Meadow:** Ibid.

195 **A recent USDA study:** "The Nationwide Microbiological Baseline Data Collection Program: Raw Chicken Parts Survey," USDA, 2012, accessed January 12, 2015, www.fsis.usda.gov/shared/PDF/Baseline_Data_Raw_Chicken_Parts.pdf.

195 **According to the CDC:** "Attribution of Foodborne Illness, 1998–2008," CDC, accessed January 12, 2015, www.cdc.gov/foodborneburden/attribution-1998-2008.html.

195 **As authors of one study:** An Pan et al., "Red Meat Consumption and Mortality: Results from Two Prospective Cohort Studies," *Archives of Internal Medicine* 172 (2012): 555–563.

196 **In the US, 80 percent:** Mindi Schneider and Shefali Sharma, "China's Pork Miracle? Agribusiness and Development in China's Pork Industry," Institute for Agriculture and Trade Policy, February 2014 (see Chendong Pi, Zhang

Rou, and Sarah Horowitz, "Fair or Fowl? Industrialization of Poultry Production in China," Institute for Agriculture and Trade Policy, February 2014).

196 **At the same time**: Ibid.

196 **Studies have found:** Alan Mathew, Robin Cissell, and S. Liamthong, "Antibiotic Resistance in Bacteria Associated with Food Animals: A United States Perspective of Livestock Production," *Foodborne Pathogens and Disease* 4 (2007): 115–133.

196 **It could be easier:** Henk Westhoek et al., "Food Choices, Health and Environment: Effects of Cutting Europe's Meat and Dairy Intake," *Global Environmental Change* 26 (2014): 196–205.

196 **In Europe, 23 percent:** Ibid.

196 **As Jeremy Rifkin wrote:** Jeremy Rifkin, *Beyond Beef: The Rise and Fall of the Cattle Culture* (New York: Penguin Books, 1992), 289 (see David Kesmodel and Laurie Burkitt, "Inside China's Supersanitary Chicken Farms," *Wall Street Journal*, December 9, 2013, accessed December 3, 2014, http://online.wsj.com/articles/SB10001424052702303559504579197662165181956).

196 **What about unemployment:** L. V. Anderson, "What If Everyone in the World Became a Vegetarian?," Slate, May 1, 2014, accessed January 12, 2015, www.slate.com/articles/health_and_science/feed_the_world/2014/05/meat_eating_and_climate_change_vegetarians_impact_on_the_economy_antibiotics.html.

196 **To take just one example:** "Meat Atlas," Friends of the Earth Europe.

197 **I'm talking about:** Ibid.

198 **Potatoes used to be:** Paul Fieldhouse, *Food and Nutrition: Customs and Culture* (Cheltenham, UK: Nelson Thornes, 1998), 56 (see "Hearing to Review the Proposal of the United States Department of Agriculture for the 2007 Farm Bill with Respect to Specialty Crops and Organic Agriculture," US Government Printing Office, accessed December 4, 2014, www.gpo.gov/fdsys/pkg/CHRG-110hhrg48113/html/CHRG-110hhrg48113.htm).

198 **Marie Antoinette wore:** Ibid.

198 **Communist fare:** E. N. Anderson, *Everyone Eats: Understanding Food and Culture* (New York: New York University Press, 2005), 106.

Index

Anka Górajka

MARTA ZARASKA is a Polish Canadian journalist who works as both a foreign affairs correspondent and a science writer. Her articles have appeared in the *Washington Post*, *Scientific American*, *The Atlantic*, the *Los Angeles Times*, *National Geographic Traveler*, the *Boston Globe*, and Poland's leading weekly newsmagazine *Polityka*, among many other publications. Zaraska is a graduate of the University of Warsaw, where she received her MA in law. She lives in France with her husband, daughter, and their two old dogs. For more information, visit www.zaraska.com.